The Power of Faith Through Spiritual Growth In Jesus Christ Our Lord

By

Bishop Everett H. Jefferson, Sr.

and

Leslie Holloway

Table of Contents

Preface

It was such a great blessing when **Bishop George W. Hawkins** came to our church, **Emmanuel Temple of the Apostolic Faith, Columbia, MD,** and preached the sermon, "**The Blessing from Bitten Hands.**" The sermon was based on the story of the Apostle Paul being shipwrecked on the island called Malta. While on the island, the Apostle Paul was gathering wood to make a fire. When he placed the wood on the fire, a viper jumped out from the fire and bit the Apostle Paul. The story of the Apostle Paul being bitten by a venomous snake, yet surviving the venous bite, is found in ***Acts 28:1-10***. Paul wasn't looking for a viper to play with so everyone could see how powerful he was. The snake was there, unnoticed, in the wood that Paul was getting ready to add to a fire. The snake jumped out from the heat of the fire and bit Paul! The people of the island concluded that Paul was a murderer who had escaped by way of the sea. However, they concluded that Paul's judgment was determined, and his punishment had come. Paul shook the poisonous snake into the fire. After a period of time, the people of the island expected Paul to fall dead from the poison of the venomous snake. Nevertheless, the people of the island noticed that Paul was not affected by the venomous snake bite. The people of the island concluded that Paul was a god! Our knowledge and understanding of the Word of the Lord is that born-again believers are not gods, but those who are filled with the Spirit of the Living God will do great things as the Word of God expresses. The Apostle Paul understood the Word of the Lord, that if he was to be bitten by a poisonous viper, he would not be affected by the venous beast. Was Paul so special that the bite of such a poisonous snake would not affect him? I don't think so. Paul was a believer in Christ and the Word of the Lord. Paul kept on working, and no harm from the venous snake bite caused him any harm. Paul remained living on the island and conducted a healing ministry while he was there. The father of Publius was sick with a fever and dysentery (infection of the intestines). Paul laid hands and prayed for Publius' father, and he was healed. The situation with the viper and Paul laying hands on the sick man, convinced the people on the island that Paul was a strong believer in the Almighty God! Paul kept on working, and the attack of the viper was of no effect. The Lord showed favor and mercy to the Apostle Paul, which amazed the people of Malta. People on the island were also healed from diseases when Paul prayed to the Lord Jesus. Consider the miraculous events that took place as the Lord Jesus used Paul to preach and teach the Gospel of salvation. The scriptures did not say whether the religion of the island changed or how many people believed or received salvation. However, I believe changes took place in the hearts of the people as many may have believed and given their lives to Christ Jesus. Was Paul greater than the average person who is filled with the Spirit because he was able to do these things? I don't think so. Paul believed and acted on the Word of the Lord Jesus. Can we, born-again believers, do the same thing? I am convinced that born-again believers have done such things and can still do such things as the Lord God Almighty allows. The Lord will work in us to do great things as we believe and act on the Word that the Lord has given. I believe that miracles still happen! Nevertheless, such things that were exhibited through the Apostle Paul and others are possible for any born-again Christian who will believe and act on the Word of God. The Bible is filled with spiritual examples of Christ and born-again believers doing great things that show the power of Jesus Christ working from within the believers. It is imperative that we have faith in the Word of the Lord and live the life that is given to us through the Holy Spirit. The spiritual work that the Apostle Paul did on the island of Malta may have influenced many to come to Christ. The scripture does not say, but I believe that life on the island of Malta changed as miracles happened and the people saw the power of God in action. When people see the power of Christ working in the born-again believer, there is curiosity as to how such things happen. I believe that such things will occur in our spiritual lives when we are obedient to the call that Almighty God has given us. I admonish every born-again believer to be obedient to every Word that the Lord gives us to speak and demonstrate the power that we have been given that will change the life of a sinner to a follower of Christ. The saints of Almighty God have been empowered with the Spirit of God that is living within them. There are great things that we can accomplish through the Holy Spirit that dwells in us. Do you believe that what the Lord spoke to His disciples applies to us also? I do, as well as many other Christians. Unfortunately, many Christians do not believe the written Word of God and miss the many blessings that the Lord has for us. Why was the Apostle Paul successful in his spiritual walk in Christ? Paul believed and acted on his belief! Let us believe what the Lord Jesus has said and is saying to us through the written and spoken Word of God. The Apostle Paul declares, ***"But we have this treasure in earthen vessels, that the excellence of the power may be of God and not of us. We are hard-pressed on every side, yet not crushed; we are perplexed, but not in despair; persecuted, but not forsaken; struck down, but not destroyed—" (2 Corinthians 4:7-9, NKJV).***

Introduction

The sermon that was preached to us by **Bishop George W. Hawkins** increased my faith and helped me to be strong in the Lord and to continue doing whatever the Lord speaks, regardless of my physical condition. Are there situations in your life that are holding you back from doing what the Lord Jesus has called you to do? This is a question to be considered, not to be ignored and forgotten. The Lord desires to use His people as a witness to all the world. If we allow the Lord Jesus to use us to do great and miraculous things, we will certainly glorify Almighty God! Do you remember the spiritual mandate the Lord gave to His disciples before He ascended into heaven? ***"But you shall receive power when the Holy Spirit has come upon you; and you shall be witnesses to Me in Jerusalem, and in all Judea and Samaria, and to the end of the earth." (Act 1:8, NKJV).*** The Lord Jesus gave specific instructions to His disciples that they would receive the power of the Holy Spirit and be witnesses throughout the earth.

During this time, there was no mass transit or rapid means of transportation.

Nevertheless, the Word of God grew and manifested throughout the earth. People were being saved as the Word of God was being taught to people all over the world. When the means of communication and travel increased, the Word of the Lord also increased. The Word of the Lord concerning salvation is being spread throughout the entire earth. Are you doing the work of bringing people to Christ as you witness the saving grace of Christ Jesus that has come into your life? I pray that you are doing this great work as people are being saved in your home, neighborhood, or wherever you are and being led to speak salvation. I am often reminded of the words of the Apostle Paul, ***"For I am not ashamed of the gospel of Christ, for it is the power of God to salvation for everyone who believes…"*** It is the Gospel that brings salvation to the world. The spiritual job of the believers is to be a witness, spreading the Word of salvation to as many as the Lord directs you to speak. The power the Lord has given us has the authority to change the lives of all who will believe. The Word that the Lord spoke to us through one or many believers changed our lives. We walked away from the old life as we began to live and commit our lives to Christ. The power of faith will grow in the child of God when we grow spiritually. Growing in grace and knowledge of the Lord Jesus is a powerful process. Many may wonder and marvel at how some children of God arrive at this status. It is the desire to learn and the sacrifice we make to know Jesus, and the power of His resurrection and the fellowship of His suffering. Jesus informed His disciples, ***"Take my yoke upon you, and learn from me, for I am gentle and lowly in heart, and you will find rest for your souls." (Matthew 11:29, ESV).*** The Lord Jesus informed His disciples to learn from Him. If you are a disciple of Christ, should you not be learning the Word of the Lord through the spiritual sources that are made available to us? During this twenty-first century, there are many spiritual resources that will help us to learn the Word of the Lord so that we may grow in grace and knowledge. Take advantage of the spiritual resources we have been privileged to receive. Our spiritual life must grow in order to fight and effectively defeat the enemy. As our knowledge of the Word increases, so will the power of faith increase. We cannot remain babes in Christ. This is a cruel world we live in. Each day, the abundance of corruption and evil grows greater. The spiritual actions we employed many years ago may be effective in some spiritual fights, but as time grows closer to the coming of the Lord, the enemy has stepped up his evil ways. The church must increase in the wisdom and power of the Lord. We cannot be afraid or timid. We must fight the good fight, and this fight requires the power of faith! I am determined to win and see the glory of our Lord Jesus Christ. We have come too far to lose this battle, and the Lord is with us! Let us continue to grow in grace, power, and authority so we may see the goodness of the Lord in our daily lives.

The Lord Jesus has given us spiritual gifts and talents to manifest His power on the earth and prove to the people that the Lord's power is with us and within every born-again believer. The Lord Jesus works through the born-again believers to show His power on the earth and cause the unbelievers to believe and give their lives to Christ. Will we, as born-again believers, be an example of the love and work of Christ? When the love and power of Christ Jesus is seen in us, others will be amazed and desire to experience the love and power working in them. The Lord desires that all people be saved and come to the knowledge of the truth ***(1 Timothy 2:4)***. Our mandate as born-again believers is to do the spiritual work that the Lord has commissioned us to do and lead people to salvation. The authority that the Lord has given us is powerful! The Holy Spirit that dwells in the believer allows us to do great things, showing the world that Almighty God exists in the baptized believers through the Name of Jesus Christ our Lord. Therefore, we have been given the authority to do great spiritual and natural things on earth and in heaven. Note the Word of God, "***For God has not given us a spirit of fear, but of power and of love and of a sound mind" (2 Timothy 1:7, NKJV).*** Our lives are changed by the Word of the Lord! Let us pursue and live by every Word that is given to us by our Lord Jesus Christ!

Chapter 1

Our First Divine Call and Action

You therefore, my son, be strong in the grace that is in Christ Jesus. And the things that you have heard from me among many witnesses, commit these to faithful men who will be able to teach others also.

(2 Timothy 2:1-2, NKJV).

In the Book of Luke, a physician writes one of the Gospels as he carefully documents the perfect humanity of the Son of Man, Jesus Christ. Luke emphasizes Jesus' ancestry, birth, and early life before moving carefully, and chronologically tells of His earthly ministry. Luke is also the writer of the book of the Acts of the Apostles. The second volume of Luke's work tells the story of the beginning of the church. It begins with the ascension of Jesus Christ and the giving of the Holy Spirit at Pentecost. Luke also discusses the life of the early church and the conversion of Saul, a Pharisee who persecuted the church and his conversion to Christianity. Luke's writing focuses on the beginnings of the church, then on the work of Peter, the growth of the newly formed church, the establishment of the deacons, and finally on the ministerial work of the Apostle Paul. Luke wanted to show how the church spread from Jerusalem and over Palestine and to the Gentiles. One of the important first mandates declared to the early church was, ***"…and you shall be witnesses to Me in Jerusalem, and in all Judea and Samaria, and to the end of the earth." (Acts 1:8b, NKJV).*** When we read the four Gospels and the Epistles in the New Testaments, it is quite clear that the first order of our spiritual work, after receiving the Holy Spirit, is to witness. I believe that we should continue telling others of the saving grace of our Lord Jesus Christ as long as we live on this earth. Before the new converts consider the official leadership in the church, the newborn children of God should express the saving power of Jesus Christ in their lives. The new converts may not be able to teach the detailed instructions of salvation, but they can certainly tell others what happened to them when they received salvation. The initial evidence of receiving salvation motivated the new believers to tell others what had happened in their lives. They were so excited about the infilling of the Holy Spirit that they were motivated to tell others about this great salvation they had experienced. The infilling of the Holy Spirit in their lives brought such great joy that the first thing they were motivated to do was to tell others about their experience and what they had just received in the baptism of the Holy Spirit. This excitement happened with the first church over two thousand years ago, and it is still happening today! It is like the excitement of the marriage of a new bride and groom, sharing the wedding ceremony with their friends and family. Oh, the joy and excitement that fills their life and motivates them to tell someone the great news that just happened to them! Such motivation occurred in my life when I was filled with the Holy Spirit of promise that changed my life. Remember the song by Bill Gaither? *"Get All Excited Go Tell Everybody That Jesus Christ is King!"* This was the first divine call in the lives of most Christians when they were called into the salvation of Jesus Christ.

Great excitement and telling others about the saving grace of Jesus Christ was normal for the newborn Christian. Our witness about the saving grace of Christ Jesus in our lives motivated many souls to receive the salvation of the Lord in their lives and tell family and friends about this great salvation. I believe as long as we embrace our salvation, we will be motivated to tell someone how the Lord has saved us from sin and unrighteousness and placed our names in heaven's eternal record. When Peter spoke to the new converts on the Day of Pentecost, and Paul spoke to the Jews and Gentiles after his conversion, they encouraged the people of God to be witnesses and walk in the calling that the Lord had committed to their lives. The instructions from Peter, Paul, and the other leaders encouraged the children of Almighty God to remain faithful to the call of Christ in their lives. The infilling of the Holy Spirit gave all new converts the power and authority to witness to the people in the world and encouraged other new converts to remain faithful in their walk in Christ Jesus. The new converts of the early church were called the Way. These new converts grew and allowed the manifestation of the Holt Spirit to be a powerful force in the earth. The Lord allowed the church to grow, and the people of this new faith spread to different parts of the world. Saul of Tarsus, a Pharisee, was one of the Jewish leaders who fought against the new converts by having many imprisoned. However, as Saul was on his way to Damascus to imprison any that were among the Way, a light shined from heaven and blinded him. Saul fell to the ground and heard a voice saying, ***"Saul, Saul, why are you persecuting Me?" And he said, "Who are You, Lord?" Then the Lord said, "I am Jesus, whom you are persecuting…" (Acts 9:4-5, NKJV).*** As Saul was trembling after hearing the message, he said, ***"Lord, what do You want me to do?"*** The Lord gave Saul instructions on what he needed to do and instructions to Ananias, who was to lead Saul to Christ Jesus. When Ananias met with Saul and ministered to him, Saul received his sight and was baptized. Immediately, as noted in the scriptures, after Saul's conversion, he preached that Christ was the Son of God in the synagogues.

The Lord works in a different way to those He calls to spread the message of salvation to the world. Saul, who they also called Paul, immediately preached Christ in the synagogues ***(Acts 9:20)***. Remember, Paul was a Pharisee well learned in the Jewish faith and traditions. The Lord used Paul to preach to Jews and Gentiles who gave their lives to the Lord when they heard the Gospel preached. Paul was not ashamed of the Gospel. He declared the Gospel to be the power of God unto salvation ***(Romans 1:16)***. I am determined to know the Gospel of Jesus Christ because it is the power that brings salvation to every person! Those of us who are enjoying this great salvation and looking forward to the return of the Lord to earth are waiting with great expectation and joy. I believe that if we remember the day of salvation and the great joy that filled our hearts, our joy will remain in full power and great expectation! Remember, it is the joy of the Lord that causes us to be strong and gives us the determination to continue fighting. ***(Nehemiah 8:10b).*** Nehemiah's encouragement to those who were rebuilding the walls that were torn down around Jerusalem is also an encouragement to all of us to keep on doing the work of the Lord. We should never give up because the Lord is with us, and He will never leave us nor forsake us.

When we read the first actions of the Old and New Testament saints, it reflects great joy and the privilege to be called by Almighty God to His work. We will experience troublesome times and sometimes great mental struggles. But if we follow the leading of the Holy Spirit in our lives, we will experience great joy, peace, and victory as we continue in the faith. Never forget the spiritual power that came into your life, the infilling of the Holy Spirit! Remember, ***it is the power of God unto salvation***. When we remain excited about the power of Almighty God living in us, others will see it and express the desire to experience the joy we are experiencing. The early church grew exponentially as the word of salvation was spread to others with great joy and excitement. Are you excited about your salvation? I am! I cannot forget the day the Lord Jesus Christ saved me from the awful state of sin that controlled my life. The awful state of sin destroyed my life and others that we closely associated with me.

Nevertheless, when the Lord Jesus brought salvation into my life, it changed me and those with whom I was associated. Let me recommend to those who have obtained this great salvation. Do not be ashamed, be confident, and be willing to spread the Gospel to as many as the Lord Jesus directs. When the Lord Jesus tells you to speak to someone about this great salvation, pray, speak what the Lord tells you, and don't hesitate, it will save their life from this world and prepare them for the Kingdom of Heaven. Your witness to others is a powerful force! It will change the life of the most dreadful sinner. Lives will be changed, and the Lord Jesus will be most pleased with your spiritual actions. Always remember, one of your first orders of business after salvation is to witness (pray, rejoice, and acknowledge what the Lord has done). Walk with the Lord Jesus each day, and He will build you up with great strength and power to do the work of witnessing that is in your heart.

I am convinced, as the scriptures have declared, ***"Being confident of this very thing, that He who has begun a good work in you will complete it until the***

day of Jesus Christ." (Philippians 1:3, NKJV). It is important for us to remember that it is the Lord Jesus Christ who initiates His work in us and will bring His work to completion as we committee our hearts to the Lord and the people He desires for us in His service. Our commitment to the Lord and people is an act of genuine love from the heart of Almighty God. Just as the Lord has committed this great Gospel to us, likewise, we should committee the Word of the Lord to faithful men and women who are able to teach others. When this is done, the Kingdom of God will remain strengthened and powerful even in the midst of a wicked and sinful world. Nevertheless, we will endure and be successful when the tricks of the enemy come against us. Let us fight and win, using the authority that is in us by the Holy Spirit. We may not win every battle, but we will win the war that comes against us. I am determined that WE SHALL OVERCOME BY THE BLOOD OF THE LAMB! Therefore, we will continue fighting!

Chapter 2

Preparing for the Gospel Work

"But grow in the grace and knowledge of our Lord and Savior Jesus Christ." (2 Peter 3:18, NKJV)

I admonish all Christians to remember the Lord's call on your life. Do not be ashamed of the salvation that you have been blessed to receive. The ultimate reward of your salvation (return of Jesus Christ) is closer than you can ever imagine. You must also remember that Christ in you is working through you so that the Word of salvation be impressed upon the hearts of others, and they will testify to others about this great salvation. We are admonished by the Word of the Lord to ***"Grow in grace and knowledge..."*** We will never be able to completely testify to others about the salvation we speak and live unless we grow in grace and in the knowledge of Christ Jesus. People are going to carefully examine the words we speak and the life we live. Our Christian character will be admired by some people who are impressed by the godly change in our life and criticized by others who are determined to destroy the godly witness that we are presenting. Whether our words are admired, examined, accepted, or rejected, it is our spiritual mandate to witness. The effectiveness of our witness will depend on our knowledge of the salvation that lives within us. As we mature in Christ, it is certain that we will meet the same people at some time in our day-to-day living. Your lifestyle in Christ will cause them to be inquisitive about our spiritual behavior. Will we tell our friends, enemies, and associates the same spiritual witness that we expressed to the members of the church initially? Will you be spiritually equipped to give them more, or will your word of testimony be the same and leave them desiring to hear more? The lead scripture for this chapter encourages the saints to grow in grace and knowledge. The Apostle Peter also declares, "***For this very reason, make every effort to supplement your faith with goodness, goodness with knowledge, knowledge with self-control, self-control with endurance, endurance with godliness, godliness with brotherly affection, and brotherly affection with love.***

For if these qualities are yours and are increasing, they will keep you from being useless or unfruitful in the knowledge of our Lord Jesus Christ. (2Peter 1:5-8, HCSB).

In our walk in the Lord and our expression of faith to others, we should add, along with our testimony, the production of the fruit of the Spirit. Faith is the spiritual foundation of our existence. There must be things that we add to our spiritual existence that will allow us to grow in the Spirit and exhibit godly qualities in the earth. During my youth in my hometown of Helena, Arkansas, when people saw me walking from school or downtown, they would ask the question, *"Are you Percy and Katie Jefferson's son"?* (my father and mother). I would say yes, and the person asking would declare, *"I thought so!"* They would continue their conversation by saying, *"You walk and talk like your daddy and look like your mama."* I was quite pleased when they acknowledged my likeness to my parents. I believe that every child of God should allow the attributes of their Heavenly Father to be active and shine throughout their lives.

When we live and exhibit the likeness of our Heavenly Father, it is greatly noticed. Some people will notice the godly qualities you possess and will appreciate and desire to understand and even obtain the likeness of your faith. Remember the basketball commercial *"Be Like Mike,"* introducing Michael Jordan in 1991? The greatness of this one-of-a-kind basketball player motivated many basketball players to strive to play like him. Michael Jordon was considered and still is the greatest basketball player of all time. I believe if we allow the character of our Lord Jesus to shine out of our spiritual being, there will be others desiring to be like Jesus because they see Christ in you!

Growing in grace and knowledge is not a difficult task if the basics of our existence are on a Solid Foundation (Jesus Christ) and we love our Heavenly Father, who allowed us to obtain new birth. When we continue to add the good building blocks of our Creators and our parents to our spiritual and natural growth, we become good spiritual and natural persons who will do great things during our existence on earth. Growing in grace and knowledge is not achieved overnight. When goodness begins to grow, knowledge will sprout up also. The nature of our public and private schools is a prime example. Our parents, grandparents, close relatives, and good neighbors laid the foundation of a good and wholesome life within us. Our parents and teachers gave us the basic knowledge to be successful in life with people, prosperity, and personality. Let us continue in spiritual development. First, let us consider the spiritual character of faith in ***2 Peter 1:5.*** The Bible defines and declares that faith is: ***"Now faith is the <u>substance</u> of things hoped for, the <u>evidence</u> of things not seen." (Hebrews 11:1, NKJV).*** Notice the two words underlined in ***Hebrews 11:1***. Let's consider two or more examples tied in with ***Hebrews 11:1***. "*Faith means relying completely on who Jesus is and what he has done to be made right with God.*" **(Harmon, Matthew S.)**, and *"Faith is a Word from God received into the heart of the believer and acted on by the power of God through the Holy Spirit of the believer."* **(Jefferson, E.H.)**. If we carefully examine ***Hebrews 11***, we will find these expressions. "*By faith: Abel offered, Noah moved, Abraham obeyed and offered his* *son Isaac, Sarah conceived, Isaac blessed Jacob, and Joseph made mention of and gave instructions."* There are many other great people of God that when their story is told, there is a Word from God and an action that follows, producing the magnificent glory of God in their lives and others! It is obvious from reading the Word of God that there was a Word from God, a positive action by those who heard from God, and the glory that followed for the benefit of generations to come. Hebrews declares, ***"And without faith it is impossible to please God…" (Hebrews 11:6a, NIV)***.

Chapter 3

The Building Boxes of Our Faith: Part 1

"But grow in the grace and knowledge of our Lord and Savior Jesus Christ." (2 Peter 3:18, NKJV)

Let us examine the first four powerful attributes that we are instructed to add to our **faith** *(goodness, knowledge, self-control, endurance.)* These four attributes contribute greatly to the child of God's spiritual growth and development from within. I will be using both the HCSB and NKJV Bibles for clarity in this illustration. These attributes are added to your faith to increase the spirituality of the child of God.

Goodness/Virtue - *Goodness, the quality of being morally good. Virtue, having or showing high moral standards.* When we obtain favor (salvation) from Almighty God, He places these divine attributes in us to grow, develop, and function as His godly offspring, being the true example of His character in Heaven and on earth. If you closely notice the offspring of humans and most animals, they are taught by their parents the basic functions of living and proper care for themselves. The Apostle Paul makes this declaration, ***"For in Him we live and move and have our being, as also some of your own poets have said, For we are also His offspring." (Acts 17:28, NKJV).*** The *goodness/virtue* that comes from Almighty God is developed in us on a godly scale through new birth (infilling of the Holy Spirit). Our growth and development in Christ Jesus allow these attributes to develop at a higher level than the time we were born again. The more we grow and develop in Christ, *goodness/virtue* becomes a platform for the other attributes to be added. The high moral standards that are placed in us through natural and spiritual birth began to grow and manifest a greater degree of glory than when we first believed. Consider a fruit or vegetable seed that is planted in the ground. When it sprouts up from the soil, it is a beautiful sight to behold. However, it is a greater sight when the fruit we are expecting appears, grows to maturity, and is served on the dinner table. Some would say, *"It's ready to be* *consumed!*" The Lord God desires to use us in His Kingdom to attract others and lead them to new birth and the Kingdom of God. Remember the words of Jesus, ***"Let your light so shine before men, that they may see your good works and glorify your Father in heaven." (Matthew 5:16, NKJV).***

Knowledge - Now, let's examine other fruits from our garden. The person who grows a garden or plants fruit trees desires to understand what it takes for the quality of fruit to be exceptional and good for consumption. Almighty God saves and places us on the earth to be witnesses to others, giving knowledge and wisdom for people to consider and obtain salvation. The next spiritual attribute that we will add to or build on to our faith and goodness is *"knowledge." Knowledge is facts, information, and skills acquired by a person through experience or education, the theoretical or practical understanding of a subject.* When we are born-again, the meaning of life and wisdom changes to a higher spiritual perspective. The facts, information, and skills we acquire may not change naturally but take on the additional characteristics of spirituality. Yes, there is the wisdom of the world and the wisdom of Almighty God. This spiritual and natural wisdom abides in the born-again believer. It helps the child of God to understand the spiritual and natural relationships to be applied during the witnessing process. When *knowledge* is inclusive in our Christian walk, the words we speak, natural and spiritual, are easily understood and accepted. We have heard the expression, *"Knowledge is Power!"* This quotation was given by Sir Frances Bacon in 1597 and continued to be used by Thomas Jefferson in connection with the establishment of a state university in Virginia. Using the expression in our spiritual life gives us focus, meaning, and determination to continue in the good work that the Lord has commissioned us to do. Mankind has acquired and used knowledge to do great things for the improvement of our lifestyles. The word *Omniscient,* associated with our Christian studies, means *all-knowing*. Of course, this theology is used to describe Almighty God. He is the only one who knows everything. Therefore, the Apostle Paul specifies to the church in Colossae, ***"...That you may be filled with the knowledge of His will in all wisdom and spiritual understanding..." (Colossians 1:9b, NKJV).*** I have come to understand that spiritual knowledge is one of the greatest accesses that the children of God can possess. When we are born, we use natural knowledge in the creative process to enhance the condition of our lives. As we grow and develop naturally, natural knowledge increases. Likewise, the same process develops in the infant or newborn child of God. Remember, ***"We grow in grace and knowledge..."*** When we grow in grace and in the knowledge of Christ, we enhance our spiritual and natural abilities. When we speak the Word of the Lord to the sinner and saint, it will evoke the desire to learn and even become a child in the Kingdom of God. Whether we are in our natural or spiritual stage, our growth gives us the maturity to live better and focus on the things that will enhance our ability to do great things as we advance in life. Consider the medical professionals who serve us so that we may maintain good health. The knowledge they obtain about the human body never ends. The career of medical professionals never ends. Our knowledge about our Lord and Savior is endless also. We will never know every detail about Almighty God. However, the more we know, the greater our life will become. The spiritual knowledge will enhance our lives greatly!

Self-control - *the ability to regulate and alter your responses to avoid undesirable behaviors, increase desirable ones, and achieve long-term goals*. There are four types of self-control: physical movement, emotion, concentration, and impulses. It is obvious that there is a measure of self-control that is within us when we are born. As we grow and develop the ability to regulate our responses in life, we become structured and successful. The natural example I have just described also applies to our spiritual development. When babies desire the necessary nutrition to satisfy the hunger cravings that arise, they cry until someone gives them the necessary nourishments to satisfy their hunger. However, when an adult becomes hungry, they do not exhibit the same action. The adult's urge to satisfy their hunger is quite different than that of an infant. The adult will reason within their mind what food they desire to satisfy their hunger. They will prepare the meal or buy it and then consume the food that they have prepared. It is reasonable to conclude that the ability to control their actions is a learned behavior. This is a simple, natural example of self-control. Let us examine a complex example that the majority of humans experience. Anger! It happens to the best or worst of us at the most inappropriate moment. We may desire something, but there is interference from loved ones, friends, and even enemies. When most human beings are opposed by others, they have strong feelings, which can lead to opposition, anger, and even fights. There are two scriptures in the fourth chapter of the book of Ephesians that directly deal with one of the greatest things that the child of God must deal with in controlling our anger. ***"Be angry, and do not sin: do not let the sun go down on your wrath, nor give place to the devil." (Ephesians 4:26, NKJV).*** Anger has a controlling power that forces some individuals to hurt others. Anger has a controlling power that moves some people to do the extreme that could result in someone losing their life or hurting someone in a most serious way. Anger will cause jealousy, hate, or other acts that could lead people to take the life of another person. Anger can produce extreme acts of violence that will endanger the lives of many people. When we become bitter, revengeful, and malicious against others, we may lose control and allow our anger to manifest to the point of trying to destroy someone. The Apostle Paul continuously instructs us about this unpleasant behavior (anger) in ***Ephesians 4:31-32. Let all bitterness, wrath, anger, clamor, and evil speaking be put away from you, with all malice. And be kind to one another, tenderhearted, forgiving one another, even as God in Christ forgave you. (Ephesians 4:31-32, NKJV).***

Did the Apostle Paul contradict himself when discussing anger in verses twenty-six and thirty-one of Ephesians four? I don't think so! In verse twenty-six, Paul realizes that humans, even after being filled with the Holy Spirit, will become angry, but to overcome the anger that we may possess, we should not sin. Let us consider what the ***(NLT) New Living Translation*** declares, "***And "don't sin by letting anger control you." (vs 26)*** and ***Get rid of all bitterness, rage, anger, harsh words, and slander, as well as all types of evil behavior." (vs 31).*** When we lived in sin and unrighteousness, sin had significant control over our lives. Salvation changed our lives, as the Voice of the Lord, living in us, helped us to overcome evil ideals and passions that could cause us and others great harm or even death. Therefore, with the Holy Spirit living in us, we have the power to not allow anger to control us. When we grow in grace and knowledge, we will put away or get rid of the anger that may control us to do evil things.

Endurance/Perseverance - Endurance - *the fact or power of enduring an unpleasant or difficult process or situation without giving way. Perseverance - doing something despite difficulty or delay in achieving success.* Endurance in the born-again believer increases as we mature in life and godliness. The things that caused us spiritual and natural aggravation during our immaturity are being reduced to a minimum, and our lives change exponentially. Spiritual endurance helps us to overcome the adversities that so often keep us from advancing naturally and spiritually. Every individual who desires to live a good and prosperous life understands that advancement in wisdom and knowledge is the two key components in growth and development. In every aspect of our lives, we have concluded that natural and spiritual advancement is necessary for success. Our world continues to change. Understanding weather patterns, our health, modes of transportation, and skills that enhance our lifestyle should never cease to improve. Mankind has never stopped learning about the complexities of the human body and the things that we come in contact with on Earth. It is apparent that during our lifetime, we will continue to learn and advance in the knowledge and wisdom that makes our lives better. We never stop learning during our lifetime as we change and experience great avenues of life, education, and, of course, spirituality. When we consider the things that we need to achieve and survive naturally, it is a great amount of information and experience. Nevertheless, we cannot neglect our existence and well-being spiritually! Our spirituality requires so much more! The average natural life expectancy of people in this world is about 72.3 years. We have been informed by the Bible that when we accept the salvation of the Lord Jesus Christ, we live forever! Each day of our life, we grow in grace and knowledge of the Lord. The knowledge we gain during this life will be used while we live on earth, naturally, and when we live spiritually in heaven.

We are dealing with unpleasant and difficult situations, but we do not give up because Christ lives within us. Also, as we do things despite the difficulty, we will achieve spiritual and natural success. This is the power of the Spirit of God working in us. Jesus refused to come down off the cross so that the people He loved would have the privilege of obtaining eternal life. Likewise, we will have a cross to bear so that someone will see our love for Christ and consider a life of salvation. This is the spiritual maturity that is acceptable before Almighty God. We will endure hardships, difficulties, and even persecution as we live godly in this life. However, the spiritual hardships and persecutions that we face cannot be compared with the glory that will be revealed in us. ***(Romans 8:18).*** Therefore, my brothers and sisters allow the attributes of our Lord Jesus Christ to grow and develop in us. The scriptures declare, ***"Be assured that the testing of your faith [through experience] produces endurance [leading to spiritual maturity, and inner peace]." (James 1:3, AMP).*** We will be tested as we grow in grace and knowledge. With each test of our faith, endurance and many other godly attributes are increasing. As these attributes increase, we become the spiritual fighting machine that Almighty God wants us to be! Our spiritual fight is "*the good fight of faith.*" When we fight, we will win the battles, the war, and experience the eternal bliss on earth, and when enter everlasting life!

Chapter 4

The Building Boxes of Our Faith: Part 2

For I am not ashamed of the gospel, because it is God's power for salvation to everyone who believes, first to the Jew, and also to the Greek. For in it God's righteousness is revealed from faith to faith, just as it is written: The righteous will live by faith. (Romans 1:16-17, HCSB)

Let us examine the last three powerful attributes that are added to our faith. (*Godliness, brotherly affection, and love*). These three attributes reflect the outward holy appearance of the Lord Jesus Christ in the born-again believer. God's righteousness is revealed through the believer for all the world to see, understand, and desire. These three attributes are expressed directly to the people we live with and meet during our life on the earth. They also confirm, outwardly, the life of Christ that lives in us. People will have a good understanding of the righteousness of Christ as they see the life of Christ in the believers.

When the children of God are being built up, the sinners will observe our godly behavior and desire to be in Christ because of our godly behavior. We cannot be ashamed of the salvation that we have received. God's salvation and power is revealed to others through our life in Christ. The expression of righteousness through our lifestyle will motivate the nonbeliever to accept the salvation of Christ living in us. Therefore, we cannot be ashamed of this great salvation we have received. We cannot hide it, but express it to everyone that the Lord leads us to tell this great salvation. Consider what the Gospel has done in your life. Someone told you about Jesus, and you received the Lord and your life and made a great change! Now it's your turn to tell someone so they will receive Christ and their life will be changed. Consider what I say, and the Lord will give you the understanding.

Godliness - *Obedient to God's commandments and seek benefit for others. Our actions come from a place of selflessness and placing the needs of others before our own. Goodness is the nature we possess as one of* *the fruits of the Spirit, reflecting God's character.* Inclusive are the characteristic synonyms: *blessed, divine, godlike, holy, and sacred.* It is interesting and preferable by divine influence that the Lord, through the Apostle Peter, included *godliness* after (*endurance* or *perseverance).* If we analyze these seven godly attributes, the function and strength of each is placed in the order that will support the next attribute that follows. When we read about the character of the Heroes of Faith (Abraham, Isaac, Jacob, Moses, and others) mentioned in ***Hebrews 11***, we find a number of specific godly qualities that gave them their heroic character. Each of the heroes of faith suffered severe adversity! Nevertheless, they conquered kingdoms and experienced mocking, scourging, stoning, imprisonment, afflictions, and mistreatment. Yet these heroes of faith obtained a good testimony because of their faith. Let us examine these heroic characteristics in a natural example during this present time. The first preseason football game played is the *Hall of Fame Game*. This first game of the football season is a time to honor a group of talented and exceptional players to be recognized for their great talent while playing in the *National Football League.* A chosen number of football players accepting this award will receive a bust (a sculpture of the football player's head, shoulders, and chest), a ring, a decorative jacket, and worldwide recognition for their accomplishments in the National Pro Football League. Each year, a great ceremony is given for these Hall of Fame player's exceptional football accomplishments. These men are celebrated as heroes because of their great athletic accomplishments. The heroes of faith in the Bible are celebrated because of their belief, faith, and actions that glorify Almighty God and change the lives of many others. One day, the Lord Jesus will return to receive His Bride (the church). We will be gathered together and celebrated for the new birth we have received and the acts of faith that we exhibited during our time on earth through Christ Jesus. Oh, what a day that will be! No longer will we suffer the corruption that now exists in this world or the wicked devices of the devil that were continually used to ensnare us. We will be free from sin and empowered to destroy the satanic influence that exists on the earth. However, before the children of God receive this great reward, there is an abundance of difficult spiritual work involving obedience to the Lord and opposition from the devil. Like the football players mentioned, there is a great fight and determination to win for our specific team. Our spiritual fight is similar, yet on a much higher spiritual order, to enter Heaven's Hall of Fame. There are struggles, fights, and even tears as we are engaged in the spiritual battles we encounter. The football player who enters the Hall of Fame is not just an average player. In order to achieve this prestigious award (induction into the Hall of Fame), endurance and perseverance are necessary to be the base that supports the award of a Hall of Famer. Our induction into Heaven's Hall of Fame requires hard work and dedication to the Word of God. Consider our test as training exercises to strengthen our spiritual muscles for the good fight of faith. These spiritual training exercises are not easy, and we may fail some of the tests. But we are determined to experience God's good hand of good that we will keep the faith, stay in training, and when we fight, we win! Why do we win? Christ Jesus lives in us. We are not alone. Jesus will assure us of victory! Remember, all the previous spiritual building blocks *(goodness, knowledge, self-control, and endurance)* are ingrained in us before we are ready to receive the last two that the child of God must have for the ultimate spiritual completeness. Therefore, my brothers and sisters, remember these spiritual fights are good fights, and we will certainly win as we prepare for the journey ahead. Keep fighting, never give up, and experience the greatness of God in your life. Many children of God are determined to fight and win the spiritual prize. Are you going to win?

Brotherly Kindness/Brotherly Affection – *Brotherly Kindness* is *an extension of the natural affection associated with near kin toward the greater community of fellow believers*. When we are filled with the Holy Spirit, we become a community of believers who express the love of God to each other and the world.

Our witness to each other promotes encouragement among our brothers and sisters and attracts and motivates others to desire the Spirit of God living within them. The church is composed of born-again believers called to spread the message of salvation to the entire world. Our lives are integrated into the Kingdom of God to do good works and promote a life of godliness and peace with each other. *Brotherly affection presents a calm atmosphere of serenity, like kinship or friendship*. If we are going to witness to the world, brotherly kindness and affection must be shown as the "*genuine article*" (*a person or thing considered to be an authentic and excellent example of their kind*). The brotherly kindness we possess is one of the greatest spiritual attractions among the believers in Christ and people in the world. Our godly behavior in the world will motivate people to give their lives to Christ, desiring to be holy on earth and the expectation of entering the Kingdom of Heaven. When the children of God have grown to this level in Christ, the world will notice and desire to be like us and, ultimately, like Christ. Remember the television commercial *"Be Like*

Mike?" This commercial attracted basketball fans and players around the world to enjoy or be like the greatest basketball player ever! The professional basketball games involving Michael Jordan attracted millions of fans all over the world! Therefore, many young basketball players wanted to be like Mike. How will anyone know how to be like Christ unless they read the Word and witness our godly behavior? Remember Jesus' Word to His disciples? ***"You are the salt of the earth…"*** and ***"You are the light of the world…" (Matthew 5:13a & 14a, NKJV).*** These Words from the Lord were not only for Jesus' disciples but for all of us who have given our lives to Christ and strive to live a holy and righteous life before the Lord and all people.

Have you ever thought about why *Brotherly Kindness/Brotherly Affection* is next to the end of the powerful attributes that we are instructed to have and must add to our **faith?** Well, it takes the combination of the first five spiritual attributes before we can be spiritually grounded to show the real love of God towards each other. Before we can show the real ***Agape***

Love and its powerful force, there must be *kindness* and *affection* towards each other! Remember the Words from our Lord Jesus, ***"You shall love the LORD your God with all your heart, with all your soul, and with all your mind. This is the first and great commandment. And the second is like it: You shall love your neighbor as yourself. On these two commandments hang all the Law and the Prophets." (Matthew 22:37-40, NKJV).*** Let's explain *Agape Love: Greek agape, in the New Testament, is the fatherly love of God for humans, as well as the human reciprocal love for God.* When Almighty God fills us with His love through His Spirit, brotherly kindness/affection becomes the base that supports the godly love (agape) that we must possess to be the shining light that the world can witness. We show this agape love when we fulfill the two commandments given by Jesus in the New Testament. Love God and Love your neighbor. We no longer try to love the old human nature before salvation. We love God, our neighbors, and ourselves with the agape love that is given to us when we are born again. Let's explain agape love in our next secession.

Love/Agape Love - *is most often crowned as the highest form of Christian love. It is the kind of love and action that shows empathy, extends the desire for the good of the beloved, wants the best, extends help, or demonstrates good intentions, and is intended for everyone.* Now, let us advance to the seventh powerful attribute we are instructed to add to our **faith.** ***Agape Love*** - unconditional "God" love; the highest form of Christian love. Agape love is the highest form of love and is contrasted with eros, or erotic love, and philia, or brotherly love. If we were to use building blocks to illustrate this point, we would start with *goodness* and add *knowledge, self-control, endurance, godliness, and brotherly affection.* Our final block topping these six blocks is *agape love,* the love of God! We understand this to be the love of God because of the Holy Spirit that dwells within us. The *agape love* that lives in us motivates and directs us to love like God because the Spirit of God is living within us. The infilling of the Holy Spirit changes our lives. Not only are we directed away from the carnal things we once desired and directed our lives to. We have the Spirit of the living God in us, and it gives us spiritual focus and determination to honor our Lord Jesus Christ and love each other. This is a great part of the new birth process. We become God-like because a part of God lives in us, the Holy Spirit. Since the born-again believer has the Holy Spirit living within us, we can do phenomenal spiritual things that will enhance the Kingdom of God on earth. Remember, our first divine purpose is to witness, telling people about the saving grace of our Lord Jesus Christ. When we reach the stage of ***"Agape Love"*** and all the other six attributes are effectively working in us, we will overcome fear and the words of discouragement that will flow from others to do us great harm! The people we speak to about the salvation of Christ Jesus will desire to be like us or, rather, the Christ in us. The people that we meet and greet will be motivated by the words of life coming from within us. When we allow the seven attributes to grow and manifest within us, the very words we speak will give life to the hearers, motivating us to be like Jesus. The words we speak will not be threatening or malicious, but words of kindness, loving, and inviting to the hearer to accept the salvation of the Lord Jesus Christ! Therefore, ***"For if these things are yours and abound, you will be neither barren nor unfruitful in the knowledge of our Lord Jesus Christ. (2 Peter 1:8, NKJV).*** Since the new birth of Christ Jesus lives within us, we have much to learn and do to bring souls into the Kingdom of God. I often consider a parable in the scriptures of the talents (5, 2, and 1) given to the master's servants. ***(Matthew 25:14-30)***. Each servant was expected to increase the number of talents that they were given. The two servants who received five and two talents doubled what they received. The servant with the one talent hid the talent he had received and gave it back to the Lord when He returned. The Lord was very unhappy with the servant who received the one talent and did nothing to increase what he was given. Please read the scripture and think about it. Our salvation is very valuable to the Lord and others. We cannot receive this great salvation and do nothing with it! Our Lord is expecting us to do something to enhance the Kingdom of God. Jesus knows what we are capable of doing to enrich our lives and others. Do not be afraid and bury your spiritual talents. The Apostle Paul explains this quite clearly, "***For God is working in you, giving you the desire and the power to do what pleases him." (Philippians 2:13, NLT***). Let's be mindful of the Lord's directions for our lives. The Lord will lead and guide us in all truths and the knowledge of His will. We must not reason with others or within ourselves that following the Lord is difficult. When the Holy Spirit is given to us, we will understand the will of the Lord for our lives. Remember what the Lord said, ***"My sheep hear my voice…"*** Therefore, hear what the Lord is saying to you. Grow in grace, knowledge, wisdom, and the power of the Lord Jesus Christ. The Lord will not allow you to fail. Remember, ***"You, dear children, are from God and have overcome them, because the one who is in you is greater than the one who is in the world." (1 John 4:4, NIV).*** When the *agape love* flows from heart to heart, people will realize that God is real and desire to embrace the salvation of Jesus Christ for themselves! Invest your time and spiritual talents to glorify our Lord Jesus Christ. Allow the agape love to manifest within you, and motivate others to desire to have this same kind of love working in them. The Lord has given us every spiritual tool we need to be effective children of God, making us powerful and effective witnesses on the earth! Do not be ashamed of the spiritual power that the Lord has placed within you. Our job as Christians is to tell others about the saving power of Christ Jesus. Our Lord will change the spiritual condition of the hearts of people. Remember, the Spirit of the Lord will grow and increase souls in the earth.

Chapter 5

Not Ashamed of the Gospel

For I am not ashamed of the gospel of Christ, for it is the power of God to salvation for everyone who believes… (Romans 1:16a, NKJV)

One of the initial steps to promoting the Gospel of Jesus Christ is to be bold and confident in your faith. If we appear ashamed or unmotivated to express the Gospel, we will automatically cause those we speak with to turn away. When the Lord Jesus saved us, He gave us the greatest spiritual power that the world has ever known. The power of God to introduce salvation to every believer. When we are given power and enthusiasm by the Lord, we are expected to express the Gospel with great excitement. We will be motivated and excited to tell others how the Lord saved us from the sin that once ruled and almost destroyed our lives. The reason that we are not ashamed is because the power of Almighty God is dwelling within. Our expression of the Gospel is not expressed in timidity.

The Lord gives boldness and courage to express this wonderful salvation to anyone we come in contact with. Why should Christians be so confident in expressing the Gospel? It is powerful and glorious! When the Spirit of the Living Savior dwells within us, stand up and speak with confidence. The Word of God that we speak as we witness to others is powerful and enacts the movement of salvation in the one who believes. The power that's within us is of Almighty God, not our own power. Nevertheless, the Lord has given us this power to speak to others in a loving manner. The loving kindness that we express to others will motivate others into the Kingdom of God. Consider this example. When a person decides to do something worthwhile and meaningful in their life, they make special preparations to become creative and professional with the skills they have acquired. Therefore, when they express and prove their skills during their presentations, the best that others will say is *"impressive!"* If a realtor is trying to sell you a house, will you buy that house if the presentation is dull, boring, and unimpressive? I don't think so! But if the realtor is energetic, knowledgeable, and outgoing, and the house is very impressive, a quick sale will be accomplished. The spiritual mandate we (children of God) have been given is to tell the story about the death, burial, and resurrection of Jesus Christ. This story that we express to others will motivate each person we witness to accept the salvation of Christ Jesus. Consider our witness to the world as the Apostle Paul makes this declaration. ***"Therefore, we are ambassadors for Christ, God making his appeal through us." (2 Corinthians 5:20, ESV).*** Your knowledge and witness of Christ will touch the hearts of many believers and motivate them to give their lives to Christ. What is an ambassador? *An ambassador is an accredited diplomat sent by a country as its official representative.* We are ambassadors for Christ, official representatives sent throughout where we live and beyond to tell others about the salvation of Jesus Christ. When we accept this mandate for our lives, we are not alone. The Lord Himself is abiding with us through the Holy Spirit! There is no need to be ashamed or timid. The power of God that's within will give us wisdom to speak and the power to attract people to embrace salvation. I am proud to be an ambassador for Christ! It gives me great joy when I tell someone about the saving grace of Christ Jesus that has happened in my life. I feel inspired, motivated, and overjoyed to express this great joy to as many as the Lord leads me to witness. Great joy will enter into our hearts as we express this great salvation to as many as the Lord directs. The Apostle Peter makes this declaration, ***"Therefore, preparing your minds for action, and being sober-minded, set your hope fully on the grace that will be brought to you at the revelation of Jesus Christ." (1 Peter 1:13, ESV).*** We have been given great work to do in the Kingdom of God! Keep moving forward and never give up. The Lord is with us. We shall overcome the actions of the enemy against us and those we witness. There is no failure in Christ Jesus. If we go forward, being directed by the Lord, we will obtain victory!

The second part of our key scripture is that the Gospel ***is the power of God to salvation for everyone who believes… (Romans 1:16a, NKJV).*** People all over the world have come to realize that there are specific mechanisms that are the driving forces behind every successful endeavor. When a car is designed and built, many parts and people (engineers and workers are included to bring the car to completion). The designers, mechanical, electrical, and automotive design engineers, together, will spend about 1500 hours to design, inspect, build, and test a car. Likewise, when a medical operation is performed, a team of doctors, anesthesiologists, nurses, and medical technicians are employed to ensure a successful operation for a patient. These medical doctors and operational assistants are trained to work together to ensure a successful operation. When we consider our life in Christ Jesus, we are also designed to work together in the Kingdom of God to bring souls into the Kingdom. The Lord has given each of His Holy Spirit believers the ability and spiritual skills to promote the Gospel. The first authoritative designations were apostles, prophets, evangelists, pastors, and teachers. ***(Ephesians 4:11)***. Spiritual gifts are the word of wisdom, the word of knowledge, faith, gifts of healing, working of miracles, prophecy, discerning of spirits, different kinds of tongues, and the interpretation of tongues, all through the same Spirit. ***(1 Corinthians 12:8-10)***. As we can see, there are plenty of spiritual authoritative individuals, ministerial positions, and individuals with spiritual gifts to help grow and edify the Body of Christ. When all these gifts and callings are working together, the Body of Christ should be working as a spiritual powerhouse for the growth and edification of the church. This spiritual concept is similar to the design and working of the automotive and professional medical individuals performing an operation. When all are working together doing their jobs, it is a sight to behold. Think about this for a moment. If Almighty God can give the natural man the ability to work together to do spectacular things in the natural (engineers and medical professionals). Surely, He can give us the spiritual ability to guide His children to do great things in the Spirit.

I cannot think of a greater and more precise fine-tuned machine than the Body of Christ working together to accomplish the will of Almighty God on earth as well as in Heaven. Remember the scripture, ***"For we are God's masterpiece. He has created us anew in Christ Jesus, so we can do the good things he planned for us long ago." (Ephesians 5:10, NLT).*** The Apostle Paul encourages the children of God to earnestly desire the best gifts, and we will be shown a more excellent way. ***(1 Corinthians 12:31).*** When the Apostle Paul made this statement, he implied that the more excellent way or the greatest gift is *love.* Love is the greatest spiritual attribute/gift of all! Nevertheless, consider this question. Are we there yet? Think about where we are in our spiritual walk. Many of the spiritual attributes and gifts during this twenty-first century were not expressed with great enthusiasm. Did God take away these spiritual attributes and gifts, or did the Body of Christ (the church) ignore or remove them from our spiritual life? I will allow you to consider this reality. Nevertheless, whether as the Christian community or individuals, many godly concepts and principles are being pushed aside because people have determined that the Christian life has become too strict and unbearable for the twenty-first-century church. We must clearly understand that the Lord does not change! ***"For I am the LORD, I do not change…" (Malachi 3:6, NKJV).*** We are living in a time when everything is changing, but we must understand the Word of the Lord has not and will not change! Do not change the Word of life that the Lord has placed in your hearts. Remain steadfast and true to the work of the Lord. Live holy and righteous before the Lord and the people in this world. People are watching those of us who profess to be genuine Christians. People are looking for the true Christian, as explained in the Bible. The spiritual mandate that the Lord has given us cannot be denied. People are waiting for the move of God from the real Christians to show them the way of salvation. Many people are wishing and hoping that someone will present real salvation to them. Many people who are looking for real salvation are searching but not finding *genuine articles*. The genuine article is the power of the Holy Spirit working effectively in the child of God. Those of us who have seriously given their lives to Christ, filled with the Spirit of God and living a godly life on the earth, will express the love and power of

God throughout the earth. Remember, the Lord is with us, always!

Living for the Lord should not only occur in a church service but throughout our daily lives. The people at church see us and exclaim that we are God's people. Nevertheless, what about the people at work, at family gatherings, and in general public and private places? Do they see the real Christian or one that is only for show and tell? The Spirit that the Lord placed in us is real, powerful, and godly! It is attractive and will motivate even the greatest sinner to be like Christ. Remember, it's not us but the power that working in us, the Holy Spirit of God. I am so glad the Lord spoke His Word and drew us into salvation. I cannot forget that day when I heard the voice of the Lord, and He gave me the desire to follow Him. Many will say as I have, *"I'm so glad that the Lord saved me!"* If you are glad, express this joy to someone that the Lord has directed you to speak. It will bless your soul and motivate you to tell someone else. Remember, the Word of the Lord is a powerful force! ***"For the word of the LORD is right, And all His work is done in***

truth. He loves righteousness and justice; The earth is full of the goodness of the LORD." (Psalms 33:4-5, NKJV). Since all of God's Word is powerful and good, let us not be ashamed of the Lord and the great things He is doing on the earth. We are blessed beyond measure, and all of us can see the goodness of the Lord throughout the earth. I'm so thankful for all the Lord is doing in the lives of His people! The Lord has not finished His blessings of grace on His people. Stay focused, diligent, and in love with the Lord and each other! The Lord is soon to return. Are you ready?

Chapter 6

Knowing the Hope of Your Calling: Part 1

There is one body and one Spirit, just as you were called in one hope of your calling; one Lord, one faith, one baptism; one God and Father of all, who is above all, and through all, and in you all. (Ephesians 4:4-6, NKJV)

The reading of ***Ephesians 4:4-6*** gives us seven more powerful qualities of expressing the oneness and precise godly characteristics that are expressed in the heart of the believer. These are *(one body, one Spirit, one hope, one Lord, one faith, one baptism, and one God/Father).* Once we grow in grace and in the knowledge of Christ Jesus, our maturity in Christ takes us to a greater spiritual level. During our next spiritual level, we began to live and effectively share with others the <u>*oneness*</u> character of Almighty God. The Lord specifically acknowledges the <u>*oneness*</u> concept in the scriptures. This one God concept given to His chosen people, Israel, is never forgotten. Consider the words of Moses, ***"Hear, O Israel: The LORD our God, the LORD is one! (Deuteronomy 6:4, NKJV)***. Throughout the Old Testament, the one God concept is expressed. The Apostle Paul begins to explain in detail to the church in Ephesus also the uniqueness of the one God concept. With the one God concept as its base, Paul expresses how to walk worthy of the call of the one God that now lives within us. Paul's letter to the church of Ephesus addresses seven concepts that we encounter in our spiritual growth: prayer, spiritual wisdom, grace through faith, unity of the faith, walking in unity, spiritual gifts, and the new man in Christ Jesus. The purpose of these seven concepts is to develop spiritual maturity so that the children of God may be able to successfully exercise their faith in Christ and lead others to salvation. Let us examine the *oneness* and precise godly characteristics that should be expressed in the heart of the believer to glorify the Lord and edify the Body of Christ.

One Body - The concept of unity describes the church as one body, with many members, not all having the same function, but interlinked as members of each other to glorify the Body of Christ and edify each other. Different gifts were given to the different members of the body or church, whereby they must serve, teach, encourage, lead, and help the other members in the Body of Christ obtain the God-given mandates to each other. ***(Romans 12:38)***. The Body of Christ is one, though it has many parts, and all the parts of the body, though many, are one body, so also is Christ. It is through the Spirit that we were all baptized into one body; whether we are of different nationalities, slaves, or free individuals, we all have been given to drink of that one Spirit. When the Apostle Paul wrote this letter to the church at Ephesus, he stressed the unity of the believers in and through Christ to exercise the faith that was in their hearts when they were born again. Consider the one-body concept as a family. We have been joined together or made alive by one Spirit. Think about that for a moment… ***"But he that is joined unto the Lord is one spirit." (1 Corinthians 6:17, KJV).*** The unity of the born-again believers in Christ is a powerful force! Think of all the things we can do together and accomplish in Christ

Jesus if we work together in the Spirit. If we observe all the great companies that manufacture the products we use, we will understand that the unity involved will bring success. In a previous chapter, I talked with my high school classmate who served as an engineer for a very large auto plant. He explained that the making of a highly efficient luxury car required hundreds of skilled workers and engineers to develop and assemble the parts necessary to manufacture a luxury car. This concept is not only in the auto industry but in many other manufacturers that produce quality products. I am convinced that the call on every believer's life is to promote the greatest spiritual product available to all mankind. That product is the production of souls in the Kingdom of God. What do we call this spiritual product? Salvation! The born-again believers in Christ Jesus are one body. When we become born-again, we have the mind of Christ. The mind of Christ must grow and develop within us so we can be the effective children of God in the Kingdom of God and on earth. When we work together, there's no limit to the spiritual and natural things we can accomplish. Let us think about this. Why are the automobile companies successful? One mind, one purpose, and one product: a great car. We accomplish our spiritual mandate when we have the oneness of Christ Jesus working in us. Our official worker's manual is the **BIBLE**! Every soul that hears our voice speaking the Gospel of Jesus Christ and is moved by the Spirit of God will respond and give their life to Christ. When we work together through the Spirit, we can accomplish anything! ***"…for you are all one in Christ Jesus." (Galatians 3:28b, NKJV).*** The accomplishments we make in the Kingdom of God are manifested when the children of God work together as a fine-tuned Holy Spirit machine. Yes, we are individuals, but the great work that we accomplish is successful when we are a unit, working together and successfully helping to develop new souls in the Kingdom of God.

One Spirit – The body is a unit, though it is made up of many parts, and though all its parts are individuals, they form one body. In order for our own bodies to function properly, there is the spirit that is within us. So, it is with Christ and the church. When we give our lives to God, He fills us with His Holy Spirit! For we were all baptized by one Spirit into one body, regardless of our race, creed, or color. ***"For by one Spirit we were all baptized into one body..." (1 Corinthians 12:13, NKJV***). We are not just members of a club or special organization; we are the Body of Christ. Individual members, yes, but a spiritual body designed to function and operate as one special, well-oiled spiritual machine! Our function: spread the word of salvation to all people, as the Lord directs, encourage and edify our brothers and sisters in the Lord, and most of all, spread the love of God to as many as the Lord directs. The one spirit concept is a powerful force. We have observed this oneness throughout our natural life, and it works! If this spiritual power is working in the Body of Christ, think of the great things that God will allow us to accomplish in our everyday lives. Since the resurrection of Jesus Christ our Lord, many signs and wonders have been recorded of the miracles that the Lord has performed through His people. Search the recorded history of these events, and you will understand how great and powerful the work of the Lord functions among the Body of Christian believers. This is why I believe and speak as the Apostle Paul, ***"For I am not ashamed of the gospel, for it is the power of God for salvation to everyone who believes..." (Romans 1:16, ESV).*** We may ask the question, what is salvation, or what are the events, signs, and wonders of salvation? Most will be knowledgeable and agree that it's the love of God. This love of God will manifest so many great signs and wonders that the Lord has done and will continue to do the things in the lives of His children. Salvation has allowed so many great things to happen in our lives that the testimonies would be endless. Therefore, to determine what are the greatest things that have happened since we obtained salvation would be an endless book or conversation. This concept makes me think about the song: *"He's Done So Much for Me I Cannot Tell It All."* We are blessed and special people who have been born-again and live in the family of Almighty God! We have the Spirit that raised Christ Jesus from the dead dwelling in us. We are a powerful people called by Almighty God to do great work on earth to enhance the Kingdom of God. Let us understand who we really are and continue to embrace the one Spirit that God has given us. The unity of the one Spirit living in us gives us the power to build the Kingdom as the Lord desires. We will experience the growth and greatness of the Kingdom of God on earth. People all around the earth will see and desire to be a part of the greatness of Almighty God on the earth!

One Hope – Hope in the New Testament means looking to the future with assurance. Because we know Jesus, we are assured of having hope. When the hope coming from God is activated in us, we can accomplish the impossible! True born-again active Christians will define, show the character of Christ Jesus, and motivate people to give their lives to Christ. When we gratefully acknowledge our salvation in Christ, people will inspect our actions to see if we exemplify the character of our Lord Jesus Christ. The visual and verbal expressions of our Christian nature will encourage some and lead others into the Kingdom of God. ***"And in His name Gentiles will trust." (Matthew 12:21, NKJV).*** When people are led to

Christ, they become hopeful of their future. What is hope? Let's consider two definitions: *(1) it is the confident affirmation that God is faithful and that He will complete what He has begun. It is also, therefore, that confident expectation which waits patiently and ardently for God's purposes to be fulfilled. (2) An optimistic state of mind that is based on an expectation of positive outcomes with respect to events and circumstances in one's life or the world at large.* Let us combine the two definitions of hope: the confident affirmation that God will complete what He has begun and an expectation of positive outcomes promised through the scriptures. Our expectation is to enter the Kingdom of God when Jesus comes to receive His church into Heaven. Is there any greater hope than this? Our one hope is being fulfilled by our new birth in Christ. Remember, Jesus paid it all! This is the one and only hope that gives us the right to the Tree of Life. Let us consider this scripture: ***"Beloved, now we are children of God; and it has not yet been revealed what we shall be, but we know that when He is revealed, we shall be like Him, for we shall see Him as He is.***

And everyone who has this hope in Him purifies himself, just as He is pure." I believe that this is the ***one hope,*** "***we shall be like Him,***" that is spoken in ***Ephesians 4:4-6.*** I look forward to that day with great expectancy! Therefore, let every child of God remain steadfast to our confession in Christ, for what God has promised will come to pass. ***(Hebrews 10:23).***

One Lord – Most English Bibles include a footnote to express alternative translations, as this is a difficult passage among Hebrew scholars. Options include ***"The LORD our God is one Lord," "The LORD is our God, the LORD is one,"*** and ***"The LORD is our God, the LORD alone."*** In all the options expressed, the focus is on the idea of one God. The doctrine of one God was in stark contrast to the theologies of the cultures surrounding the Israelites. Other religious systems served a wide variety of gods and goddesses. The worship of only one God made the faith of the Hebrews unique. The one God or one Lord concept was engrained in the Old Testament. However, when the New Testament (New Covenant) was introduced, many ideas and doctrines were introduced by man, a trinitarian (three Gods), and multi-god concepts were introduced throughout the world. The concept of one God did not diminish when the New Covenant was explained by Jesus and the apostles. When the people were filled with the Spirit of God (Holy Spirit), great things began to happen, but the original one God concept never changed. They understood just as the Old Testament and Jesus also specified, ***"The LORD our God is one Lord."*** Think about this for a moment. Why would the Creator change the one God concept when He has already declared that the Lord is One! Jesus Christ was God incarnated in the flesh to die on the cross and save mankind. Let us examine the scriptures, Old and New Testaments, ***"Therefore My own arm brought salvation for Me; And My own fury, it sustained Me." (Isaiah 63:5b, NKJV).*** And ***"In the beginning was the Word, and the Word was with God, and the Word was God." (John 1:1, NKJV).*** Ladies and gentlemen, let us not be confused! The Lord is One and always will be One. He does not change His authority or power. He is One God and Lord over everything, forever and always! Remember, the Lord did not change the one God concept.

Let us explore. The origin of monotheism is indicated in the opening words of the Bible: ***"In the beginning, God created the heavens and the earth." (Genesis 1:1).*** Only one God was before all things and created all things. This same one God was the One who spoke with Adam and Eve in the Garden of Eden, saved the world through Noah, and promised Abraham that all nations would be blessed through him. Israel was always taught that the Lord God was the one God. The nation of Israel was informed to reject all forms of polytheism. The Father, Jesus Christ, and the Holy Spirit are each referred to as God, and many Christians believe that God is one! We know God through faith in Jesus Christ ***(John 14:6)***, and the Holy Spirit works within us to help us live for God each day. Even though we may use these titles independently, there is only One Lord and Father of us all!

Chapter 7

Knowing the Hope of Your Calling: Part 2

Therefore we also pray always for you that our God would count you worthy of this calling, and fulfill all the good pleasure of His goodness and the work of faith with power… (2 Thessalians 1:11, NKJV)

One Faith – In addition to one Lord, there is also *"one faith"* ***(Ephesians 4:5)***, an expression with two possible interpretations: the body of content that is believed (as, for example, in Jude 3) or the subjective experience of faith that all believers have as they trust in Christ alone for redemption. Part of this appeal for Christians is to strive, ***"to keep the unity of the Spirit in the bond of peace" (Ephesians 4:3).*** The Lord God of heaven has given us *"one faith."* Not multiple faiths, but one designed to bring unity to the Body of Christ. Faith is from the Greek word, which can refer to one's personal faith, one's faithfulness, or that which is called *"the faith."* While it is true that there is only one Savior, who is the Mediator between God and man, the meaning of *"one faith"* is not clear to everyone. Well, we do not all believe the same thing when it comes to the doctrines that mankind concludes of the Bible. Some believe that the Father, Son, and Holy Spirit are three separate persons. Nevertheless, when the Apostle Paul expressed the *oneness* concept in ***Ephesians 4:4-6*** (*one body, one Spirit, one hope, one Lord, one faith, one baptism, one God and Father),* greater clarity was expressed to explain the unique character of Almighty God! When we observe what is recorded in the Bible, it is clear that the one God concept has not changed. If the oneness of God does not change, why will the characteristics of God and the people that are born of His Spirit change to the multi-god concept. I may be out on a limb to some, as I constantly reiterate that the *Lord is One*. Nevertheless, as we constantly read the Word of the Lord and pray for the wisdom of God to become clear to all believers, we will clearly understand that there is only One God. I believe that the *"one faith"* is the Gospel of Jesus Christ or the Words of Christ that led us to salvation. During the time of the Pharisees and Saul of Tarsus, there was a serious attempt to destroy *"the faith"* ***(Galatians 1:23)***. After Saul's conversion, what did he do? He preached "*the faith*". Yet, as Paul described his role, he would say that he preached *"the gospel"* ***(Romans 1:15)***. Again, "one faith" is the same as *"the faith,"* which is the Gospel system of believing that Jesus Christ came into the world to save sinners. Here is great news for mankind: the *"one faith"* is unchanging. While some cry for a "new" gospel, something that is more "up-to-date," we need to accept this reality: the Lord is not going to give us anything different from *"the faith"* that was revealed by God's power and wisdom. The Gospel's message is, ***"the faith which was once for all delivered to the saints" (Jude 3)***. The *"one faith"* is compared to a seed, an incorruptible seed that lives and abides forever ***(1 Peter 1:23-25)***. The *"one faith"* includes facts that humans must believe in order to be saved ***(1 Corinthians 15:3-4)***. It includes commands from God that must be obeyed ***(Hebrews 5:9)***. *"The faith"* also includes warnings to be heeded ***(Matthew 10:28)***, as well as promises of blessings to be received ***(Acts 2:38)***. In short, *"one faith"* is the message that men need for the salvation of their souls . . . it is all-sufficient for every human need. Note some words/terms associated with the *"one faith."* It is truth ***(John 17:17)***, it is power ***(Romans 1:16),*** it is life ***(John 6:63,68),*** and it is light ***(2 Corinthians 4:4)***. With God as our Helper, let us keep on studying, living, proclaiming, and defending the *one faith.* This *one faith* will allow the children of God to grow and manifest the power of the Kingdom of God on earth. Therefore, ***Let us hold fast the profession of our faith without wavering; (for he is faithful that promised;) Hebrews 10:23, KJV).***

<u>**One Baptism**</u> – ***Ephesians 4:4*** also says, "There is ***one baptism***." Since there are different "baptisms" referred to in the New Testament, it can be a bit confusing when we read about *"one baptism."* The word *baptize* always *means "to submerge or immerse."* When water baptism is discussed, it involves a person being totally submerged in water. Baptism implies being *"all in."* It also implies that a change has taken place. Baptized people are declared to be changed people.

Generally speaking, there are two types of baptism: a

physical (water) baptism and a spiritual baptism. One is done in water; the other is accomplished through the infilling of the Holy Spirit. Water was commanded by Jesus for all of His followers ***(Acts 1:5-8)***. ***Colossians 2:12*** says, ***"Having been buried with him in baptism, in which you were also raised with him through faith in the powerful working of the Holy Spirit, who raised him from the dead."*** Being baptized with water does not save us; faith in the finished work of Christ and the infilling of the Holy Spirit saves us! The things that I am discussing in this section *(one baptism*) can be complicated if we do not cover a plethora of scriptures. Let us consider a few more scriptures. ***"And it shall come to pass that whoever calls on the name of the LORD shall be saved.*" *(Acts 2:21, NKJV); "Believe on the Lord Jesus Christ, and you will be saved…" (Acts 16:31); and "Moreover, brethren, I declare to you the gospel which I preached to you, which also you received and in which you stand, by which also you are saved, if you hold fast that word which I preached to you—unless you believed in vain. (1 Corinthians 15:1, NKJV).*** Nevertheless, water baptism is an outward indication of an inward change. It is a wonderful picture of Christ's death, burial, and resurrection. Being immersed in water symbolizes the cleansing of our hearts and the washing away of our sins by the blood of Jesus ***(Acts 2:38)***. Through water baptism, believers publicly proclaim that they have entered the second phase of the new birth process. The first phase is repentance, the second is baptism in water, and the third is the receiving of the Holy Spirit ***(Acts 2:38)***. ***Romans 6:3*** speaks of a spiritual baptism: ***"Or don't you know that all of us who were baptized into Christ Jesus were baptized into his death?"*** This spiritual baptism *"into Christ"* is performed by the Holy Spirit the moment a repentant sinner accepts the gift of salvation and is born again. ***Jesus answered, "Most assuredly, I say to you, unless one is born of water <u>and the Spirit</u>, he cannot enter the kingdom of God. (John 3:5); "For through Him we both have access by one Spirit to the Father. (Ephesians 2:18); "For by one Spirit we were all baptized into one body." (1 Corinthians 12:13a)***. We respond to the Holy Spirit's drawing us to salvation, and we are born into God's family. By this "baptism," we are identified with the death and resurrection of Jesus; from then on, we consider ourselves "crucified with Christ" ***(Galatians 2:20)***. We choose to lose ourselves and be immersed in Christ. The baptism of the Holy Spirit was promised by John the Baptist, who said, "***Jesus will baptize you with the Holy Spirit and with fire." (Luke 3:16)***. No one understood what John the Baptist meant until after Jesus had ascended back into heaven ***(Acts 1:9)***. Jesus had promised the disciples that He would send *"the Comforter"* ***(John 14:26; 15:26; and Luke 24:49)***. His followers were to wait in Jerusalem until the *"promise from the Father"* came ***(Acts 1:4)***. That promise came in ***Acts 2***. The Holy Spirit was poured out on the disciples, and many received the Holy Spirit and were never the same again. They were bold in their witness, empowered to perform miracles, willing to endure persecution, and all but one died a martyr's death. The church had begun, and great signs and wonders were manifested on the earth. Throughout the book of ***Acts***, that baptism by the Holy Spirit was repeated as people came to know Jesus, both Jew and

Gentile and served to unify the church as the Jewish believers realized that the Holy Spirit was poured out on the Gentile believers as well. There are some differences of opinion among believers concerning the baptism of the Spirit. Some Christians believe Holy Spirit baptism is identical to being baptized into Christ, and that occurs at the moment of salvation, even if the believer is unconscious of it. Other Christians believe that Holy Spirit baptism is to be equated with the filling of the Spirit, and that often occurs after we believe and as the believer opens himself up to the Spirit's control. Some believe that the baptism of the Spirit is always accompanied by signs, such as speaking in tongues, and others believe that such signs are unnecessary. When Paul wrote to the Ephesian believers about *"one baptism,"* he was reminding them that, regardless of their background or nationality, they all served the same Lord, shared the same faith, and had experienced the same baptism. He could be referring to water baptism, i.e., all believers have the same testimony of salvation and have been baptized in water in the name of the Father, Son, and Holy Spirit ***(Matthew 28:19)***, or in Jesus Name ***(Acts 2:38)***. I believe Paul was referring to Spirit baptism, i.e., all believers have been placed into the Body of Christ through the Spirit's power. Either way, the emphasis is on unity among Christians. The Holy Spirit works to unify believers and provides assurance that they are children of God. ***(Romans 8:16 and Ephesians 1:13-14***). We are reminded that the members of the Lord's church all had a similar testimony and that they were all partakers in the same Holy Spirit. We are encouraged by the apostles to work together for the cause of Christ so that the message of redemption would continue to spread throughout the world. Remember, we are joined together in Christ by the Holy Spirit that dwells in each of us. When we are filled or baptized with the Holy Spirit, it is undeniable that we are sons of God!

One God/Father – Having explained the last six powerful qualities of expressing the oneness and precise godly characteristics, let us complete the circuit as we talk about the One God/Father, as stated in chapter six and other chapters in this book. We have discussed the one God concept indicated throughout the scriptures as ***"The Lord is One!"*** Throughout the Bible, the One God concept is reiterated often to make sure the believers in Christ Jesus understand that there is only one real authentic ***God!*** Since the beginning of time, we understand through written history and Biblical manuscripts that mankind had the propensity to acknowledge and serve multiple gods. The *One God* concept began when the Lord called Abram. God changed Abram's name to Abraham and called him the father of faith and the father of many nations. The *One God* concept was instilled into Abraham's heart. The scriptures declare, ***"…Abraham believed God, and it was accounted to him for righteousness."*** (***Romans 4:3, NKJV).*** The Lord God called Abraham and entered into a covenant relationship with him. Abraham's whole life changed as he began to acknowledge God and entered into this covenant relationship with God. Abraham is given the promise that he and his descendants will become a blessing to all humanity. God spoke with Abraham, telling him he would become the father of many nations. Also, he would become exceedingly fruitful, and kings would come out from him. God established His everlasting covenant with Abraham and his descendants. The call of Abraham was the beginning of the monotheistic *(One God)* concept to all mankind.

The descendants of Abraham grew and manifested throughout the southeast regions of Asia and other parts of the world. The first major offspring of Abraham were Isaac, Jacob, and Jacob's twelve sons. The monotheistic principle grew and manifested throughout the whole world. When the Israelites dissented into Egypt, they became slaves and stayed in captivity for four hundred and thirty years. The children of Israel grew to a great number of people while in Egypt. After their many years as slaves in Egypt, they cried to Almighty God for deliverance. The Lord God heard their cry and sent a man called Moses to request the leader of Egypt (Pharaoh) to set the children of Israel free. After many signs and wonders, Pharaoh felt compelled to release the children of Israel from bondage. Great signs and wonders were manifested before and after Israel left Egypt. These miraculous events are still discussed through writings and moving pictures of Israel's exodus from slavery in Egypt. Throughout the birth of Israel until this day, the *One God* concept remained effective. When Jesus was born, through the Abrahamic covenant, the *One God* concept was still powerful and will not be changed. In the Bible, from Genesis to Revelation, we can find the same concept, spoken with authority and power, that ***"The Lord is One!"*** Even though we may be surrounded by many documents, preaching, and discussions about the *polytheistic principles* of God, it will never override the spiritual fact that God is one! When Jesus came into this world, He emphasized the facts, ***"I and my Father are one." (John 10:30, NKJV)*** and "***For there are three that bear witness in heaven: the Father, the Word, and the Holy Spirit; and these three are one." (1 John 5:7, NKJV).*** Most Christians understand that what we receive from Bible teachings and written publications may not give us a complete understanding of the oneness of God. Many Bible scholars have tried to completely understand the three-in-one concept as discussed in the New Testament. Many that have a good understanding of the *One God* concept are rooted and grounded in the Old and New Testament. I believe it is more than just the academics of the written Word but also the God-given revelation of Biblical truths. One scripture that helped me to understand the One God concept is, ***"Hear, O Israel: The LORD our God, the LORD is one!" (Deuteronomy 6:4, NKJV)*** and ***Jesus answered him, "The first of all the commandments is: Hear, O Israel, the LORD our God, the LORD is one." (Mark 12:29, NKJV).*** We have two scriptures, one from the Old and New Testaments, declaring the *One God* concept. I am convinced and will continue to teach and preach the oneness of God as it is presented in the Bible. The majority of the world's religions teach the polytheistic principle. Although those who believe in the monotheistic principle are few, I will hold on to the written Word and the revelation that has been given by Almighty God. "He Lord thy God is one!"

Chapter 8

Living by Faith

I have been crucified with Christ; it is no longer I who live, but Christ lives in me; and the life which I now live in the flesh I live by faith in the Son of God, who loved me and gave Himself for me. (Galatians 2:20, NKJV).

There is one scripture in the Old Testament and three scriptures in the New Testament that make this declaration, ***"The just shall live by faith." (Habakkuk 2:4; Romans 1:17; Galatians 3:11; and Hebrews 10:38).*** The information we have received from the previous seven chapters gives the child of God the necessary spiritual tools to enhance our journey towards the Kingdom of God. The component of faith must be inclusive in our daily walk in Christ. If the child of God is lacking faith, they cannot please God! Sounds like a demanding behavior for the children of God from what we may expect from a loving Savior who gave His life to save us. Demanding, yes, but also necessary to please the Almighty God that saved us. Through our Christian walk in Christ, we have heard a variety of thoughts and interpretations about the Biblical meaning of faith. In chapter two of this book, three explanations were given for faith. We have made this conclusion that faith is: (1) ***"Faith is the substance of things hoped for, the evidence of things not seen." (Hebrews 11:1).*** (2) *"Faith means relying completely on who Jesus is and what He has done for us to be made right with God."* **(Harmon, Matthew S.)**, and (3) *"Faith is a Word from God received into the heart of the believer and acted on by the power of God through the Holy Spirit of the believer.* The first definition is directly from the Bible. The last two definitions are interpretations from clergy based on ***Hebrews 11:1***. When we examine the word faith a little closer, it becomes clearer what faith is. One reference to faith concludes, *"The central feature of faith is confidence or trust."* Think about this for a moment… Every Biblical definition of faith and the actions of those that the Lord determined that have faith involved hearing God speak and positive action from the hearer when the Word was received and acted upon. Example: Abraham heard God's instructions to offer his son Isaac as a burnt offering unto the Lord. Abraham acted on the Word that God gave him, binding his son Isaac to the altar of sacrifice and raising his knife to kill his son. A lamb was discovered and used as the offering of sacrifice. ***(Genesis 22)***. Likewise, Moses was instructed by Almighty God to open the Red Sea with the stretching of his hands, holding the rod God had given him, and the nation of Israel would cross over the Red Sea on dry land. ***(Exodus 14)***. The concept of faith is specified by others such as Sarah, Jacob, Joseph, Rahab, Samson, David, and others determined to hear God and act on the Word that God had given. Notice what each of these individuals did what was necessary that determined their reward from Almighty God and a place in Biblical history. They heard from God, acted on what they heard, and God performed miracles through them. Each case presented in ***Hebrews 11***, and the true testimonies during the time of Jesus until the present, requires a Word from God and the action of those who received God's Word.

Let's move to another example of faith that should be functional in the heart of all believers. ***"But do you want to know, O foolish man, that faith without works is dead?" (James 2:20, NKJV)***. Every person in the Bible, and even those living today, have received a Word from God, and the performance of works that follow will allow them to accomplish what the Lord had spoken. Let's use a scientific example. There are two types of energy, as explained in most science books: kinetic and potential energy. Kinetic energy is active energy. This kind of energy is always in motion. The other kind of energy is potential energy, energy that is stored up, non-active. Let's use a simple flashlight as an example. When the switch on the flashlight is turned off, the batteries and the light bulb in the flashlight are at rest, with no flow of direct current (DC). Therefore, the light bulb will not give light. When the switch is turned on, the flashlight is activated, and the bulb and batteries produce light. When the switch is in the on position, the batteries are energized, the direct current flows to the lightbulb from the batteries, and the light appears. When there is no flow of direct current, the flashlight does not give light. The bulb and the batteries are at rest. The flashlight and all its components are at rest; therefore, *potential energy* is in effect. Upon turning the switch on, the current from the batteries produces electrical energy that allows the bulb to shine, giving light to its surroundings. The energy that is released from the batteries and causes light to be produced is called *kinetic energy,* and the flashlight is fully operational! When these two major components are at rest, there is no light. When these components are in motion, light appears. This is a crude illustration, but I hope this natural example is effective in explaining the Apostle Peter's explanation, ***"Faith without works is dead!"*** Let's bring it home. The Lord speaks to His children through His written Word (Bible) or by His voice into the hearts of His children. If we receive the Word spoken to us in our hearts and never act upon it, there is no activity or results. If we hear the Word of the Lord and act on the spoken or written Word, we will obtain results, just as the flashlight in the "on" position! This is why the Lord Jesus declared, ***"Let your light so***

shine before men, that they may see your good works and glorify your Father in heaven." Just think about it: when our light is shining, we are producing spiritual kinetic energy. When the Word of God in the believer is in action, others will see, hear, feel, taste, and touch the power of God in our lives. The Lord God has given His children *"real/active faith."* Children of God act on the real faith that God has given us. Many people believe what the Word of God says but will not act on it. We cannot function as real children of God if we do not hear and respond to what the Lord is telling us. When we hear a word from God, it may conflict with what we think as human beings. Nevertheless, it is imperative that we hear and respond to the voice of the Lord rather than responding to human thoughts and actions that are not consistent with the Word of God.

During the latter years of the nineteen-nineties, the Lord clearly spoke to me twice over a three-year period, telling me to leave St. Louis, MO, and come to the Washington DC area and start a church. My family, friends, and coworkers informed me that I was being totally irrational. I had a great family, home, and an outstanding job. Why would I leave such a wonderful blessing behind? Why would I leave a great church and great friends and go to a place that's more expensive to live in and has no friends? I explain to my family, friends, and coworkers that I heard from the Lord. Well, none of them accepted what I said. Nevertheless, I knew it was the voice of the Lord I heard. I got the largest moving truck available, loaded it with most of my household things, and drove from St. Louis to Laurel, MD. The job was challenging. The people were different, and the traffic in the Washington, D.C. area was terrible! Nevertheless, my wife and youngest son moved into a glorious home, which was a miracle from God, and the job became better with each week. We celebrated our arrival in Maryland and never looked back. We missed St. Louis, but this was the promise from God. So many great things happened while we settled in Maryland, and after two years, the Lord spoke concerning the starting of a church. I was concerned about the physical move but terrified about the spiritual move. Nevertheless, it worked as the Lord instructed, and the church has grown and is still growing, as the Lord declared.

I have a greater understanding of faith these days. A word of faith from God is not a guess nor just a good feeling; it is a Word from the Lord! When we hear and act on what the Word the Lord has spoken to us, we will experience spiritual and natural success. Remember, ***the just shall live by his faith***… It is faith all the time and not just some of the time. The Lord has great things in store for His people. However, children of God must be attentive and listen to the voice of the Lord so that we may receive the promises He has declared for us. Consider this scripture, ***"…By which have been given to us exceedingly great and precious promises, that through these you may be partakers of the divine nature…" (2 Peter 1:3a, NKJV).*** If you desire to see the wonders of God's work in your life, faith is the main spiritual ingredient. Hear what the Lord is saying and act on what He has said. You will not fail! Remember, ***"Without faith it is impossible to please God."*** This has great things in store for us, but we must have the active faith to receive the blessings.

Our new spiritual birth has given us spiritual purpose in this life and the reward of entering into the Kingdom of Heaven for all eternity. We have been given the spiritual tools to energize the world with the Word of God that gives everlasting life to all that will answer God's call. The Apostle Paul declared in ***Galatians 2:20*** that after being crucified with Christ, we (human beings) no longer live in the human nature of our Father Adam. But we are living by faith in our Savoir Jesus Christ, who loved us and gave His life so that we might live forever with Him. What an awesome and powerful gift we have received from the Lord! In order to see and experience this great day, we must walk by faith, not by sight. Remember, faith is a Word from God. What we see (our human thoughts and perceptions) is the response to our natural reasoning. Remember, we have been born again. ***Beloved, now we are children of God; and it has not yet been revealed what we shall be, but we know that when He is revealed, we shall be like Him, for we shall see Him as He is. (1 John 3:2, NKJV)***. What an awesome promise that we are waiting to receive! However, there are many great promises that are being revealed during our life that we are now living. The Lord has so much in store for us. However, faith (active faith) must be fully functional for us to receive the promises of God in this life. The Lord desires to bless His people with every good and perfect gift. Our job is to believe and act on the Word that the Lord has given us.

Let's take our confidence in the Lord to the next level. Let us not stay at the lowest level of our spiritual walk in Christ. The Lord desires to bless His people with great spiritual and natural gifts if we only believe and act on the Word that He is speaking to us. The Apostle Paul makes this declaration, ***"I have been crucified with Christ; it is no longer I who live, but Christ lives in me; and the life which I now live in the flesh I live by faith in the Son of God, who loved me and gave Himself for me." (Galatians 2:20, NKJV).*** When we study the Word of God, we gain insight into the promises of God. Faith grows in our hearts, and we become the spiritual children of Almighty God. Faith keeps us firmly planted on solid ground. We are no longer tossed around and suffering spiritual and natural defeat in our lives. The Lord is with us, and when we hear the voice of the Lord, we will not remain in a non-active spiritual position. The Lord has so much in store for His people. Our opportunity to receive from the Lord is determined by our active faith. I believe that when God chooses us to act on a Word that He has given us, it is not for us only. Sometimes, it's for a multitude of people, from generation to generation. Your faith in the Lord may establish a new and powerful generation in Christ. Our children and generations following will give their lives to the Lord and experience great spiritual and natural things that will be continuous from one generation to another. Keep living by faith, walk in faith, and see the greatness of God in your lives. The Lord has given His children all the necessary blessings for us to have a great life in faith. We will experience adversity, but the faith we have in Christ will allow us to achieve spiritual and natural success! I'm not giving up, and nor will I toss my faith aside or allow my faith to become inactive. I desire to see the goodness of the Lord in the land that we are living in. Therefore, I will not give up. I will keep the faith and see the glory of God throughout my life here on earth!

Chapter 9

The Good Fight of Faith

Fight the good fight of faith, lay hold on eternal life, to which you were also called and have confessed the good confession in the presence of many witnesses. (1 Timothy 6:12, NKJV).

On March 8, 1971, the sports experts declared that the fight between Joe Frazier and Muhammad Ali was the greatest of all time. The professional boxing federation never before had two undefeated champions clashed for the undisputed world heavyweight boxing title. Both boxers were in their prime, and thus, strictly from a competitive standpoint, it was a dream fight! A huge attraction viewed this battle. Millions of people watched with great anticipation for the announcement of the winner! I chose to use this example of one of the greatest fights in the professional boxing world. It is still listed in the history books as the greatest boxing match thus far. I used this story to illustrate the overwhelming success that the news media achieved and the excitement that boxing fans experienced when these great fighters engaged in this great sporting event. It's been over fifty years since this fight occurred. Yet it is still declared to be one of the greatest fights of the century. This fight was viewed by many people and declared as one of the greatest of all time! Allow me to make a crude analogy between this great fight and the greatest spiritual fight of all time: Jesus, the Lord God Almighty, and Satan, aka Lucifer, the fallen archangel. Millions, if not trillions, have heard about this story through reading the Bible and Biblical stories told throughout many generations. We are experiencing this ongoing war until the last few chapters of the Book of Revelation, when Satan and all his followers are cast into eternal damnation. As of now, this fight continues... The intensity of this fight becomes great as the end-time draws near. When we were selected by Almighty God to become His children, the fight intensified! The enemy is doing everything he can to destroy human beings, especially those who have committed their lives to Christ. From the time we were born-again until Jesus comes to receive His church, we will be fighting the good fight of faith. Even after we have been caught up to meet the Lord in the air, we will fight with our Lord Jesus Christ against the satanic force that remains on the earth.

Let us begin this spiritual discussion with the Word of the Lord. ***"Beloved, now we are children of God; and it has not yet been revealed what we shall be, but we know that when He is revealed, we shall be like Him, for we shall see Him as He is." (1 John 3:2, NKJV).*** This scripture gives the children of God one of the greatest promises ever, ***"we shall be like Him and see Him."*** This is one of the great hopes that is given to the born-again believer. This is what we hope for while here on year and battling the satanic forces trying to turn us away from God. Regardless of these battles, we are declared winners by Almighty God! When our heart is fixed, and our mind is made up, we will fight until victory is declared. We will remain in this fight until Jesus returns and takes His church out of the earth. The church, the Body of Christ, has been assured of this victory if we hold on and never give up. The church is assured of this victory if we recognize the leader of this fight, the Captain of our salvation, Jesus Christ! We have been given the keys to the Kingdom and promised to win the victory when the Lord is leading the fight. Allow the Lord to take charge, and we will win the fight. Let us consider a few Biblical examples and testimonies during my lifetime. Let us not forget the miraculous things God did for His people who believed and trusted the Word that was given. Jehoshaphat, the king of Judah, was attacked by three powerful nations: Ammon, Moab, and Mount Seri. These nations were more powerful than the nation of Judah; nevertheless, Jehoshaphat prayed. The Lord instructed him to put the praises on the front line of the battle. When the praises began to say, "Praise the LORD, For His mercy endures forever," the enemies were defeated! ***(2 Chronicles 20:20-24).***

When we make a good confession in the presence of all people, we will fight a good fight. Think about the great athletes for a moment. The champion trophies begin when they are competing in exercise and bodybuilding. The true athlete completes the necessary training before entering into the real competition.

Successful training will result in great winning! Fighting against the enemy requires intense training. If we ignore the basic and intense training, we will not be ready for the war. We have been given the instruction manual (Bible) to learn spiritual warfare. The Lord has already given us victory, but we must train, know the tricks of the enemy, and stay rooted and grounded in the faith in order to win. What does the Word say? "***No weapon formed against you shall prosper, And every tongue which rises against you in judgment You shall condemn." (Isaiah 54:17a, NKJV).*** We have been declared winners before we engage in the fight! Let us consider the final outcome of the fight that the Apostle Paul was getting ready to enter. The Apostle Paul was on his way to Jerusalem when the Prophet Agabus took Paul's belt and bound Paul's hand and feet and said that the owner of this belt (Paul) would be bound and delivered into the hands of the Gentiles. When the people heard the prophecy of Agabus, they begged Paul not to go. Paul explained to them that he was going to go and was bound to die in the name of the Lord Jesus. Therefore, when Paul was determined to go to Jerusalem and die, they reframed from trying to convince him and declared, ***"The will of the Lord be done." (Acts 21:10-14).***

The prophecy of the prophet Agabus and the determination of the Apostle Paul convinced the people that he was going to Jerusalem and deliver the message that God had given him. If we are in the midst of adversity and words of prophecy that could bring us severe hardship, would you be willing to make such a journey? I believe we would if we were rooted and grounded in the faith and hearing what the Lord had informed us to do. Paul was not persuaded by the adversity that he would face but the Word he had received from the Lord. What determines spiritual success in your life? Is it the hearing of faith and a successful outcome or trying to protect your natural outcome? The hearing of faith and the action that followers will guarantee our spiritual success. Paul was tested and tried for his godly determination to enter Jerusalem. He was bound and imprisoned for the Word he had received from the Lord. Paul was spiritually successful by his obedience to the Word of the Lord! He survived adversity while being in Jerusalem, other cities, and countries, as well as many other trials for the glory of Christ Jesus. This is why Paul could speak with confidence and authority, ***"Yes, and all who desire to live godly in Christ Jesus will suffer persecution." (2 Timothy 3:12, NKJV).*** Living godly in Christ is not only believing but being active, promoting the Word of the Lord, speaking and doing things that promote the Gospel of Jesus Christ. This is the good fight of faith. We hear from God and respond to the Word He gives us to actively give to others. We cannot be ashamed. We must remember the words of the Apostle Paul, ***"For I am not ashamed of the gospel of Christ, for it is the power of God to salvation for everyone who believes... (Romans 1:16a, NKJV)***. Take courage, my friends, as we work diligently in the Kingdom of God. Fight each battle in Christ, knowing that you will win and see the glory of the Lord. I've made up my mind to fight and win! We have come too far to turn back. We are on our way to the place that God has promised. Let's consider the initial scripture ***((1 Timothy 6:12)*** of this chapter. ***"Fight the good***

fight of faith, lay hold on eternal life." Why is this a good fight? We fight by faith (the Word of the Lord). Each time we fight and win, we get closer to our divine goal, Heaven. This is why it is called a good fight. If we fight, we will win. We cannot back away from the fight and expect to be called champions. The example I used at the beginning of this chapter was the boxing championship between two of the greatest boxers during that time. Millions of people watched and were elated by the sportsmanship and beauty of that fight. What if one or both fighters did not want to complete it? That would have been the greatest disappointment of the century. However, it was not a great disappointment but a great fight! Sports fans witness the greatest fight of the century. What if the fight was canceled? There would be great disappointment and revenue loss that was inconceivable. However, the fight went forward, and many people were happy, and lots of money was made. Our fight against the enemy of our souls continues. If we back down and walk away, the great cloud of witnesses and the angels in heaven will be disappointed. But there will be no disappointment. We (the saints of Almighty God) will fight and win! With the Lord on our side and faith in our heart, we will win the greatest fight of our life! When will you declare your victory in the Lord Jesus Christ and glorify the power of Almighty God dwelling in us. Therefore, keep fighting, my brothers and sisters. It will not be long, and we shall see the great glory of our Lord Jesus Christ!

Chapter 10

The Power of Faith

And my speech and my preaching were not with persuasive words of human wisdom, but in demonstration of the Spirit and of power, that your faith should not be in the wisdom of men but in the power of God. (1 Corinthians 2:4-5, NKJV).

Our Christian world is bombarded with all kinds of theatrics and gimmicks in order to sell the Gospel of Jesus Christ. I do not wish to sound indifferent about the state of human affairs that are taking place in the world of religion. However, it is true that some of the religious cultures and practices are becoming extravagant and confusing to many people who wish to live a godly life. I do not want to infer any negative behavior that is going on in the Christian world. Nevertheless, if we observe closely what is going on, we will understand that the behavior that is happening among Christianity and other religious organizations has become confusing. When we observe and consider the things that are happening in many religions and Christian denominations, we become disenchanted and desire to walk away from our faith that meant so much when we first believed. Please do not become alarmed! These things have been happening for thousands of years in every business and religious affiliation since before the time of Christ. When these religious originations take advantage of people, many turn away and stop believing in God. Generations of people are lost because many refuse to believe in God anymore or embrace strange religious cultures that have nothing to do with the Christian culture of the Bible. This is a sad commentary for those who wish to obtain the salvation their forefathers cherished and thought to be special. When this happens, many people turn to meaningless cults and satanic practices. Many people who believe in religion desire to obtain something special and meaningful in their lives. I believe it is accurate to say that most people believe that religion plays an important role in their lives. When we read about the spiritual things that will enhance the quality of our lives, we submit our time, talents, and money to ensure that these things become desired and successful in our lives. We embrace these spiritual concepts and believe that such things will improve our lives on earth and gain us entrance to a place called Heaven. However, as we continue to practice our specific faith or religion, are we confident that our spiritual needs are being achieved? Do we see the rewards of our religious culture, or are we just performing a religious practice?

Let's consider the subject of this book and the title of this chapter, ***"The Power of Faith."*** Faith is a unique word in the Bible and in the world that we live in. Let's consider the definition of faith from the Bible and from two leaders of Christian churches *(chapter 2).* The attempt to define faith was necessary to get a clearer understanding of how it is used in our Christian lifestyle. Nevertheless, many theologians and clergy have concluded that faith is a Word from God imparted into the heart of the believer to carry out the spiritual work that God desires of His children. It was also determined in the Bible by the Apostle Peter that "***faith without works is dead***." This means that both (faith and works) must function together to be effective in obtaining the desired results. Overall, faith works in the heart of the believers to accomplish the desired tasks that Almighty God wants to achieve through His people. We are not only special people of God waiting to enter the Kingdom of God. We are also the children of God who have been given special spiritual tasks to be used to accomplish God's plan of salvation for all mankind. Our work is not finished when we believe and receive the baptism of the Holy Spirit. There are many people who need to hear about the salvation of Jesus Christ and submit their lives to the Lord. When we give our lives to Christ, this is just the beginning. As we grow in grace and knowledge, there is much spiritual word to be done. Allow your spiritual mind to be open to the voice of the Lord. Prepare yourselves to be used by the Lord, and you will be amazed at the great things that He will accomplish in your life.

Sometimes, in our attempts to do great works for the Lord, we place the thinking of the carnal man over the divine purpose of Almighty God. I believe that many Christians have discovered that this worldly practice does not work. That's why the Apostle Paul emphasized in our key scripture that his teaching and preaching did not consist of persuasive words of human wisdom. The miraculous thing that happened during Paul's ministry was the activity of the Spirit and power! God allowed this to happen so the people of God would understand that their faith was in the power of God and not in the wisdom of men. In Christian worship, we often give praise to a person for the great and miraculous things that occur in our church services. But if we are Bible students possessing the knowledge of God, we know exactly what is happening. It is the power of Almighty God working through a human vessel to accomplish these great events. Whether it is powerful teaching, preaching, healing, or deliverance, it is the Lord Himself working through a willing person to accomplish only what God can do! Wow, I feel honored to be used by the Lord to do great things in His Kingdom.

Therefore, I will not allow my flesh to interfere and receive any glory for the miraculous work that's being done to represent the Kingdom of God and the power of Christ Jesus our Lord. Consider this scripture, ***"For***

we are God's handiwork, created in Christ Jesus to do good works, which God prepared in advance for us to do." (Ephesians 2:10, NIV). We cannot do anything unless the Lord gives us the authority to do so. The power of Almighty God is real, and the working of His power in us is very effective! The more we see the great work of our Lord Jesus Christ, the more we understand that it is the Lord working through us to accomplish such great tasks. The authority we possess in Christ is powerful and effective! I am so glad that He uses His chosen people to do great and powerful things on earth. However, there are more great things to come when we are caught up to meet the Lord in the air and enter the Kingdom of Heaven! Just think, we are operating as sons of God on earth clothed in human flesh. Nevertheless, we are doing great spiritual things through the Spirit of God (Holy Spirit) dwelling in us, a person, part spiritual and part natural. But the day will come when we shall be changed. The sickness we encounter will be gone. The troubles we face will be nonexistent. All this is possible because we have been blessed by our Lord

Jesus Christ to receive the Spirit of the Living God within us! When I see the things that the Lord is doing through His people, I request the Lord to use me even more. We do not have to work at it. All we need to do is have faith. Listen to what the Lord is saying to us, and do it! We will experience great and awesome things through our life in Christ. People will see the greatness of God working in us and desire to be saved. This is why we sing the song *"How Great Thou Art."* The greatness of God is seen in the heaven above and on the earth. When we examine the great and wonderful things the Lord has done and yet is still doing, we cannot help but give the Lord praise for the things He is doing.

The power of faith dwelling in us should not be underestimated. It has been stated once, yet it is important that we continue to reiterate ***"And without faith it is impossible to please God…" (Hebrews 11:6a, NIV)***. In my opinion, faith must be at the top of our spiritual list, or love and faith are tied for first place. My opinion, what's yours? When we understand the spiritual power that we have and the glory of

Almighty God that flows from the Holy Spirit, we understand why the power of faith is so important. Every spiritual act, from believing to great acts of courage, requires the power of faith. Don't try to live without faith if you are a child of God! Consider these examples: (1) how would a doctor function efficiently without a stethoscope? (2) How would a carpenter build a house without a hammer, saw, and other construction tools? (3) How would a minister of the Gospel preach and teach the Word of God without the knowledge of the Bible? Last but not least, (4) how can we enter the Kingdom of God unless we have been born of the water and Spirit? ***(John 3:5).*** The plan of salvation that's outlined in the Gospels and the Book of Acts is not complicated. Jesus has allowed people to enter into the Kingdom of God without complications or difficulty. The plan of salvation is not difficult. The finished work of salvation was accomplished by our Lord Jesus Christ at Calvary! When the Lord Jesus speaks to us about salvation, it is pleasant and reassuring that our lives will change and we will live with the Lord for all eternity. Every action we take in

Christ requires the power of faith, from new birth to the coming of the Lord Jesus Christ. Remember, *"Without faith, we cannot please God."* The power of faith is our greatest spiritual ally. We cannot live a successful spiritual life without faith.

I admonish every spirit-filled born-again Christian to walk, live, and fight the good fight of faith that is given to us through our Lord Jesus Christ. We will experience great success as we walk and live by faith in the Lord our God. Our spiritual and natural life is complete when we exercise our faith in the Lord. Brothers and sisters, we have great confidence in our Lord Jesus Christ! The Word that has been given to us gives us the courage and power to do all things that the Lord has permitted us to do. We do not need to be afraid when threatened by adversity or any other demonic force that comes against us. We are winners! The power of faith abides in us through our Lord Jesus Christ. The enemy's weapons against us will not be effective. The power of faith assures the child of God absolute victory! Remember, we have the Spirit of God living in us. Since the Spirit is living in us, we can search for the deep things of God. When the new birth enters into us, we know the thoughts of God because the Spirit of God lives within us. I know this may sound complicated to some, but keep believing and trusting in the Holy Spirit that lives within. The Lord God has placed His Holy Spirit within each of us who believes and accepts salvation to be a powerful force in Heaven and on earth. Our work will continue throughout eternity! Therefore, beloved children of God, fight the good fight of faith and see the glory of God working effectively in our lives.

When we read the entire chapter of ***1 Corinthians chapter 2***, we will understand the quality and action of faith better. Our understanding of faith will give us the power and authority to overcome the power of the enemy. Remember, we are overcomers; we fight the good fight of faith, and we win! The things that we learn by the Spirit give us understanding and power to defeat any ungodly force. I am convinced that we are winners! We will overcome all things as we walk in Christ and fight with the power and authority we have received through the Holy Spirit of promise. Do not turn away from adversarial conditions. Fight the good fight and obtain your victory. Jesus has won the war! Keep following His Word, and victory will be ours!

Chapter 11

Spiritual Fighting with Confidence

"Therefore do not cast away your confidence, which has great reward. For you have need of endurance, so that after you have done the will of God, you may receive the promise:" (Hebrews 10:35-36, NKJV).

In chapter three of this book, we define and explain the word *"endurance."* Our lead scripture for this chapter ***(Hebrews 10:36)*** includes the word *"endurance."* Let's include another definition for the word *endurance. "The ability to withstand hardship or adversity and the ability to sustain a prolonged stressful effort or activity."* The reason that the word *endurance* stands out in this scripture and the following narrative is because of its necessity for the child of God to overcome adversity. Endurance is a necessary spiritual trait for the child of God to receive the heavenly and earthly blessings that we have been promised. Spiritual fights are necessary if we are going to see the promises of Almighty God being fulfilled during our lifetime. Our confidence in the Lord cannot be neglected or destroyed. Confidence, fighting, and endurance are necessary for the child of God to receive what God has promised for us. Sometimes, we feel that all we need to do is pray, and things will be taken care of. I agree that prayer is a part of our offensive stance. However, our fight is between offense and defense. *An offensive operation is aimed at destroying or defeating the enemy and imposing our will on the enemy for a decisive victory. Defensive operations defeat an enemy attack, buy time, economize forces, or develop conditions favorable for a counteroffensive that regains the initiative and defeats the enemy.* The child of God must be operating in the offense and defense during our war against the enemy. Both operations are required in order for the soldier to be successful. Yes, we enjoy the songs, the testimonies of the children of God, and the hearing of the Word of the Lord. All these spiritual things are necessary to build up our faith and walk on this earth. Are we in church just to hear the songs, testimonies, and preaching of the Gospel? No, we are in church to learn and develop godly characteristics to become good and effective soldiers!

Whenever there is a fight in the military, fierce competition in athletics, or any form of mental and physical competition, there is an intense strategy incorporated to win the war. We engage in connective offensive and defensive maneuvers in order to win the battle or competition. These maneuvers used in military or athletic competitions require mental and physical strength in order to win the battle. Every soldier, athlete, or someone involved in any mental and physical competition must be spiritually equipped to win the battle. One of the first and main ingredients is *confidence. Confidence is the feeling or belief that one can rely on someone or something.* The true warrior or athletes must first have *confidence* in their trainer and in themselves. The writer of ***Hebrews*** informs the child of God that our *confidence* will bring us great rewards. Our spiritual battles are meaningful to us and each other. When we engage in spiritual warfare, it is important that we win! I have decided after fifty-one years in Christ that it is important that we plan, practice, and persevere to win every battle we encounter. How should we win? In the natural and spiritual battle, it is very important to strike the first blow, continue to win each fight, and keep the enemy on the run. The real soldier or athlete realizes you must strike the first hard blow, teaching the enemy that you are in this battle to fight and win. Remember when Jesus had completed His fast of forty days and nights and was driven into the wilderness to be tempted by the devil? The devil came to Him and tempted Him with three requests. Nevertheless, before the devil made his request to Jesus, he made an alarming statement, ***"If you are the Son God…?" (Matthew 4:3).*** Such temptations will be presented before we engage in a spiritual battle with the enemy. This kind of question is designed to make us think and doubt if we are not powerful enough to entangle with the enemy. However, the negative statements that were introduced by the devil had no effect on the Lord. Jesus knew what He had been called to do and the power that resided in Him. It is very important that we first know who we are and what the Lord has sent us to do. Let us consider this event. When Jesus was baptized, He came up out of the water, and a voice of Heaven spoke and declared, ***"This is My beloved Son, in whom I am well pleased." (Matthew 3:17, NKJV)***. After Jesus' baptism, He was led into the wilderness by the Spirit to be tempted by the devil. Some theologians will declare that Jesus was tempted three times, but if you read closely, there were four temptations: (1) If you are the Son of God; (2) Turn stones into bread; (3) Throw yourself down from the highest point of the temple; and (4) All these things I will give You if You fall down and worship me. ***(Matthew 4:1-11).***

Jesus won these battles because He knew who He was and the power that was given to Him by Almighty God. The Lord has called us into His divine military and has given us the power and authority to fight and win! The devil left Jesus alone because the Lord Jesus resisted him with the Word of the Lord. The Apostle James gives us the formula for success, ***"Therefore submit to God. Resist the devil and he will flee from you. (James 4:8, NKJV).*** Resist means: *withstand the action or effect of; and to fight against or oppose* *something or someone.* Jesus fought this fight against the devil and won decisively! We who are baptized and filled with the Holy Spirit are given the spiritual power and fighting strength to overcome anything ungodly or any negative force that comes against us. Jesus won that battle, and throughout His time on earth, He won many more battles. However, when He died and rose from the dead, He won the war! Now, with the Holy Spirit living in us, we will win many battles and eventually win the war when the Lord Jesus returns to earth and takes His children to glory! We obtain our rewards through the *confidence* we have in Christ Jesus. The spiritual fights we are engaged in are already decided because we have *confidence* in what the Lord has spoken. We anticipate great rewards because the Lord cannot lie!

Nevertheless, we must have the spiritual stamina to keep fighting and have the *confidence* that we will win. One of the major factors that ensures our victory is doing the will of the Lord. When the will of the Lord has been done, we will receive the promise that the Lord has specified. Obedience and service to Almighty

God must be at the top of our spiritual bucket list. Not because we are so great but because the Lord who saved and sanctified us lives in us and has given us the authority to continue living, overcoming, and being victorious! I am determined to win because I realize I must fight the good fight of faith in Christ Jesus. Every great warrior or athlete is determined to win! In the minds of most warriors and athletes, winning is the only option. A military warrior realizes that winning the battle means they will return home to their families and receive the honor from the country they serve. The athlete is looking forward to the prize and celebrating the victory with family and friends. The children of God will celebrate our spiritual victories over the enemy of our souls and rejoice for all eternity. I am determined to win because the Lord Jesus Christ has already won the battle! Remember, Christ lives within us, and we live in Christ. When we fight, we will win, and the name of the Lord will be glorified! Nevertheless, we must grow in grace and in the knowledge of Jesus Christ our Lord. The grace from the Lord is the knowledge and discipline we receive through the Word of God that will determine the magnitude of our victory. Take note of this scripture written by the Apostle Paul, ***"All athletes are disciplined in their training. They do it to win a prize that will fade away, but we do it for an eternal prize. So I run with purpose in every step. I am not just shadowboxing. I discipline my body like an athlete, training it to do what it should. Otherwise, I fear that after preaching to others I myself might be disqualified." (1 Corinthians 9:25-27, NKJV)***. I am not expecting disappointment or disqualification. When I live my life in Christ, I expect to win and receive the prize from the Lord! With each step I take or each spiritual mile I run, I have a purpose, and my purpose is to win! I realize I will be challenged by the enemy, but I also realize that I am being encouraged and helped by our Lord Jesus Christ. Remember, brothers and sisters, we fight to obtain the heavenly prize: eternal life. Stay focused on the spoken and written Word. In doing so, we will win and obtain the heavenly prize and rejoice through all eternity!

Let us not forget that faith is a powerful ally for the child of God! We cannot overcome the forces of darkness without faith. Yes, we fight to win and receive the crown of eternal life. We fight the enemy, not only for ourselves but for all that are in the Kingdom of God. The church, the Body of Christ, lives and fights every day for each other. Remember, we are the Body of Christ! We live, work, and fight together so that we all may obtain a crown of righteousness. The scriptures declare that we are one body with many members. We win spiritual battles when we pray and fight together. We are a unique body of spiritual believers. We fight together and win together. Every born-again believer will obtain victory if we work and fight together in spirit and truth. The enemy has already been defeated! Do not give up, people of God. We have already obtained victory through Jesus Christ, our Lord! The reward of Heaven awaits us. Nevertheless, the Lord has great plans for us on earth. How do we achieve these great things while on the earth? Remember the words of the Angel that spoke to Zerubbabel, ***"Not by might nor by power, but by My***

Spirit, Says the LORD of hosts." (Zechariah 4:6, NKJV). The great spiritual things that we may achieve in this life are not by our own power but by the Spirit of the Lord! Make sure that we put our total trust in the Lord, and we will see great and miraculous things happen in this life and the life that is to come. Always remember ***Zechariah 4:6*** when we desire to see and accomplish great spiritual things happen in our life. It is the goodness of the Lord and His manifested presence that wins every battle!

Chapter 12

Endure Temptation and Receive Your Crown

Blessed is the man who endures temptation; for when he has been approved, he will receive the crown of life which the Lord has promised to those who love Him. (James 1:12, NKJV).

Throughout chapters one and eleven, we have been given specific Bible details about faith, enduring temptation, and receiving a crown for the struggles we face as children of God. As we continue reading chapter one of the Book of James, we realize that we should not say we are being tempted by Almighty God. The Lord does not tempt any of His children. But temptation is developed when we are tempted by our own desires. When the desire is conceived in our hearts, it allows sin to develop, and the development of sin brings death. ***(James 1:15-16)***. The Lord did not bring salvation into our lives so that we may obtain death. The Lord saved us and gave us life! When the

Lord filled us with His Holy Spirit, our lives changed. Now that we understand the spiritual purpose for our life, we endeavor to live spiritually regardless of the temptations and fights that come into our lives. Just like the skilled warrior or great athlete who desires to win, we know that the Lord is giving us the power to win. Therefore, we fight with great intensity, and we ultimately win! One major bit of spiritual advice I will give to every child of God is to put your trust in Jesus and rely on His Word, and we will obtain victory over the enemy. Many times, when we become seasoned or older Christians, we feel that we can accomplish great things through our own power, education, and abilities. It is absolutely untrue! Remember, the Lord Jesus is the Author and Finisher of our faith. We cannot achieve any spiritual success without the Lord. When great things happen in our lives, we may exclaim, "*I do not believe it!*" If this is our natural thinking, we will fail every time. However, we do not need to fail. The Lord lives within us and gives us the power to overcome anything that comes against us.

I am totally convinced that the Lord called me into His Kingdom, and great things are being given to us on earth, as well as the promise of a Heavenly home. The Word of the Lord has informed us that temptations will come, and our flesh sometimes desires to be carried away by these temptations. But we have been made to understand that the flesh cannot please God. ***(Romans 8:8)***. Once we have been born again, our desire is not to please our flesh, the enemy of our souls, the devil, but to please Almighty God! Think about this for a moment. Every good and wise parent will not send their child to the first day of school without careful instructions, properly dressed, and well nourished. Every parent who loves their children will do as much as they can to ensure the proper treatment and educational instructions for their child. The good parent demands respect for their children and insists that their child is treated properly. However, there comes a time when the parent must allow the child to go to school by themselves. The child has reached a specific age that the parent and child desire to have a certain independence from each other. Maturity helps the child and parent develop educationally, naturally, and spiritually. Now, our precious child begins to grow and manifest into a proud upperclassman who can handle almost everything. Our Lord Jesus loves us very much and desires that we are treated with respect and honor because we are His prize possessions whom He loves and cares for. Good parents determine when a baby is conceived the things that they desire to be a great blessing to their child. The good parent plans for their child's future. This is real love as parents make sure that their child is well taken care of and ready for the positive and negative occurrences that happen in their life. We feel proud of the accomplishments that have been developed in our children, and we feel that they are ready to do great things in their lives. However, when we look back at the success that was accomplished with the first child, we will apply the same or more instructions to the next child. The idea, we never stop learning!

We are children of God, and the Lord God makes sure we are well taken care of. The Lord wants the best for all His children. Therefore, the Lord gives us the best that we may develop into great children of the Kingdom. Consider this scripture, ***"If you then, being evil, know how to give good gifts to your children, how much more will your Father who is in heaven give good things to those who ask Him!" (Matthew 7:11, NKJV)***. Our Father in Heaven loves and cares for us. He will always be there for His children. We have no need to worry about the day-to-day events that we may encounter. As we grow and develop spiritually, we will encounter good and bad things in our lives. Some things we will prepare for, and others will take us by surprise. Just as good parents will train their children to handle good and bad natural situations, the Lord is giving us a great quality education in order that we may become good stewards in His Kingdom! Notice what the Lord proclaimed, ***"Let your conduct be without covetousness; be content with such things as you have. For He Himself has said, "I will never leave you nor forsake you." So we may boldly say: "The LORD is my helper; I will not fear. What can man do to me?" (Hebrews 13:5-6, NKJV).***

Children of God, we should not be intimated by the devil or people who are driven by their fleshly desires to hurt us. We are special, and we belong to the Kingdom of Almighty God. Who can convict us of things that make us depressed and ready to walk away from the Lord? Adversities of the spirit and flesh will try but cannot accomplish their wicked goals that will try to lead us to become fugitives of the grace of God. We must not forget that we have been empowered by the Holy Spirit living within us. Challenges will face the child of God often in order to draw us away from the love of God. There are often reminders that try to charge us with the old behavior that once existed in our sinful life. Nevertheless, if we remain in Christ, we maintain our confidence in the Lord and never go back to our former sinful lifestyle. The scriptures also inform us that, ***"Yet in all these things we are more than conquerors through Him who loved us." (Romans 8:37, NKJV)***.

It is a spiritual fact that the enemy of our souls, the devil, wants us to return under his control so that he may have another chance to destroy us. However, I hope every child of God plans to maintain their spiritual integrity and fight the enemy! When we endure temptation, the Lord builds a wall of protection around us. It is the determination of every child of God to know the Word of God and use it effectively against the enemy of our souls. Children of God, you have been given authority over the devil. You have been declared a winner in your spiritual fighting. We have come too close to our spiritual goals to give up. I've seen too much and experienced the power of the Almighty working in our lives. An old Gospel song would declare, *"I don't feel no ways tried; I've come too far from where I started. Nobody told me that the road would be easy. I don't believe He brought me this far to leave me!"* I believe that the old Gospel song speaks to the heart of many believers. The Lord did not call us into His kingdom to leave us to be recaptured by the enemy. Remember, the Lord put His Holy Spirit in us to be spiritual warriors and winners. Each day, we may encounter evil things that come to make us turn away from God. Nevertheless, there is a great power living within us that carries us to victory! Let us consider the spiritual dress that helps us overcome the enemy.

The Lord God has placed on His children the whole armor of God that is necessary for each of us to fight and win! We stand steady and sure with the *belt of truth* around our waste. This belt of truth stands out to the enemy that truth will always win out over a lie, and if you come with lies as your weapon, the truth will win every time. The *breastplate of righteousness,* protecting us from the enemy's arrows and wicked darts, are meant to pierce our bodies and damage our heart and lungs to take our strength and eventually bring spiritual death. Our feet are enclosed and ready to move with the power that comes fitted with the readiness that comes from the Gospel of peace. When our *feet are fitted with the Gospel of peace,* we are empowered to run into the battle and not away from it. The *shield of faith*, which we can use to extinguish all the flaming arrows of the devil, stands out from us as a protective force. This shield will protect the sons of God from the incoming arrows and flying objects that try to damage our entire being. And last but not the least, we will put on the *helmet of salvation* and the *sword of the Spirit*. The helmet stands out as the protector of the control center for the soldier, the mind. All fighting intel comes from the control center, and this very valuable piece of equipment (the mind) must be protected at all times. Our offensive and defensive weapon is the sword of the Spirit. We protect with our sword, and we destroy with our weapon, the sword of the Spirit. This is the Word of the Lord! ***(Ephesians 6:13-18)***. When the enemy seeks to destroy us, make sure he observes us from head to toe. Upon the enemy's attack, he will witness our helmet of salvation. Our mind is covered by the Blood of Christ Jesus. The way we think is the power and authority of the Lord Jesus. The enemy realizes that he is a defeated foe. He cannot win by going against the Lord Jesus Christ. The devil lost at Calvary when he felt assured that he had defeated the Lord God Almighty. But after the Lord's death at Calvary and His burial, there was the most powerful spiritual event ever! Christ rose from the dead just as He said before His crucifixion! Now, the enemy of our souls knows he has been defeated and will face eternal damnation. Yet while he waits for his eternal destruction, he is trying to bring as many souls as possible with him to his final destination, the lake of fire.

Therefore, beloved saints, realize with no doubt in your minds that the overcoming power within us is greater than you will ever imagine in this life. Just like the soldiers and athletes we have described, it is imperative that we stay armed and ready for battle! Do not give ourselves over to the enemy. Remember, we are winners! Remember the old saying, "*winners never quit, and losers never win.*" Many years ago, I preached a sermon entitled, *"I Am Not Defeated!"* I was determined to inform the people of the church that we are not and will not be defeated as long as we are committed to our Lord Jesus Christ! Throughout my youthful life, I played every sport I could. I was determined to win in the sport that I played very well and the sports that I just learned. I didn't like losing. As I grew older, about sixty years of age, I refrained from playing hard-hitting sports and began to coach. I loved winning even when I could not play anymore.

This same intensity followed me throughout my career in chemistry and the preaching of the Gospel. When the Lord called me to preach the Gospel in 1976, I made up my mind I was going to stay focused on the Lord. I was terrified yet determined to allow the Lord to work in me His Word and to bless as many people as I could. Now, as I live in my seventies, there is still some spiritual fighting left in me. Soon, I will approach the eighties, and may I declare as the Apostle Paul, ***"For I am now ready to be offered, and the time of my departure is at hand. I have fought a good fight, I have finished my course, I have kept the faith…" (2 Timothy 4:6-7, NKJV).*** This may sound like my departure from Earth into Heaven, but only the Lord knows. Therefore, I will continue to do the work that the Lord is calling me to do with all my heart. I am experiencing just as many spiritual fights as I encountered in my youth. Nevertheless, my spirit is strong, and I am determined to win! Regardless of your age or the amount of time you have been with the Lord, are you determined to win also? The time I spent as an athlete, I was determined to win! I refused to give up and lose to my adversity! If I lost, I would consider the reasons that caused me to lose. I would develop strategies so the next time I played, I would correct the errors that caused me to lose so I could win! Keep fighting, my brothers and sisters! Stay vigilant, focused, and determined to win this spiritual fight. Never give us, and you will see the glory of the Lord! I am determined. What about you?

Summary

Sometimes, I wonder how my life in Christ has come to this point. I was taught about the saving grace of our Lord Jesus at an early age. However, when I grew into my teenage years, I fell away from the Lord and began to do many unrighteous things. As my life began to spiral completely out of control, I heard a voice speaking to me about my evil ways and actions that were slowly consuming my life. When I heard this voice, I was not intoxicated or drugged. My mind was clear and sharp, and I was expecting a great evening with friends. I heard this voice, and the reason I wanted to understand what was happening to me was the words I heard were absolutely true! After that event, I was still living out of control, and my lifestyle was taking me further from reality into a place that would bring me heartache, sickness, and death. I began to survey my lifestyle and realized that I was out of control and my life would soon hit rock bottom. The Lord spoke to me again, and I realized that the Lord, whom I learned as a child, was telling me to come back to the place where I first believed. This was the first divine call for salvation to come back into the wonderful, good life I once enjoyed. I believe many born-again Christians may have had similar experiences that moved them into salvation. These dreams, visions, and revelations were the Lord Jesus calling us to receive the salvation He prepared for us through His death on the cross of Calvary. Thank God we heard the call and responded by submitting our lives to the Lord Jesus, getting baptized, and being filled with the Holy Spirit. Thank God we heard the call and submitted our lives to the Lord Jesus Christ.

After a period of time during my conversion, the Lord spoke again. This time, when the Lord spoke, He called me to salvation. What a great day that was. After the call for salvation, there was a time for us to prepare the witnessing and maybe even the preaching and teaching of the Gospel. Our preparation time included praying, studying the scriptures, and overcoming the sinful spirit that once ruled our lives. The Lord allowed us to grow and manifest His Spirit that we received when we were born again. The work has just begun! We were endowed with the Spirit of God, which gave us wisdom, understanding, and power to live in this sinful world. Great things began to happen to each child of God when we decided to commit our lives to the building of the Kingdom of God! This building process started when we began to understand and allow the Lord to work in each Christian the seven building blocks of the believer's faith: *goodness, knowledge, self-control, endurance, godliness, brotherly affection, and love.* These first four attributes reflect the outward holy appearance of the Lord Jesus Christ in the born-again believer. The last three attributes reflect the inward power, authority, and holy appearance of the Lord Jesus Christ in the born-again believer. God's righteousness is revealed through the believer for all the world to see, understand, and desire. These three attributes are expressed directly to the people we live with and meet during our life on the earth. All seven attributes confirm, outwardly, the life of Christ that lives in us. We can easily express these seven attributes because we are not ashamed of the salvation we have received from the Lord. God has changed our lives, and we cannot hide it from all the world. We stand firm on what we believe, and we are willing to tell everyone how the Lord saved us from sin. I am proud and grateful for the salvation I have received! Are you?

During the growth process of each child of God, we begin to deepen their knowledge of the oneness of Almighty God in our lives (***Ephesians 4:4-6).*** We are informed that there is *one body, one spirit, one hope, one Lord, one faith, one baptism, and one God*. The Apostle Paul, the author of the book of ***Ephesians***, was declared to be the gifted Apostle of the Lord with wisdom and knowledge of the salvation of Jesus Christ that we have accepted and desire to live. We notice the expression and characterization of the oneness that defines the development of the church. Remember, we are the church! Nevertheless, due to the polytheistic belief that is still common in the world, Paul emphasizes to the Body of Christ the oneness principles that are deemed critical to the believers in the Body of Christ. It is very important that all born-again believers understand that God is One and the oneness of His character extends through every aspect of His creation and Godhead. The Apostle Paul expresses that the church should live, understand, and express the life of the Oneness of God throughout our lives in Christ. The oneness of God is a power attribute in every born-again believer. When we express this oneness with the hope of our calling, we become a powerful force on the earth!

Three acts of faith are expressed in chapters eight, nine, and ten. Living by faith, the good fight of faith, and the power of faith are the main spiritual concepts that build the structure of Christian living. The scriptures admonish the believer that, "***Without faith it is impossible to please God!***" Faith is a Word from God that is ingrained in our hearts so that we may live a healthy spiritual life on this earth. Remember in the Book of Genesis when God breathed into the first man (Adam), and he became a living soul or being? ***(Genesis 2:7)***. The last Adam (Jesus) became a life-giving spirit. ***(1 Corinthians 15:45).*** This spiritual life that we were given has blessed us to live in the Kingdom of God and do great things because the Spirit of God lives in us! Our new birth experience gives us the power to overcome the evil tactics of the devil. We also live in harmony with the Lord, Christian people, and even unbelievers. Even the things we accomplish in life are given to us by our faith in the Lord. Remember Paul's words, ***"The just shall live by faith" (Romans 1:17b, NKJV).*** We cannot live this Christian life and achieve spiritual blessings without faith in God. Faith is the driving force behind the success of every child of God! When we live by faith, the sky is the limit as to what the Lord will do for His children. When we become rooted and grounded in Christ Jesus, we can do anything we have been allowed by the Holy Spirit. The Word of the Lord has declared that we fight the good fight of faith! I explained earlier that the fight we become engaged in is a good fight because we are winners. By faith, the Lord instills in the believer the power to fight and win. When we fight the good fight of faith, we experience amazing things taking place in our lives. We experience the power of Almighty God intervening in our spiritual and natural battles, giving us the victory that we can only achieve through Christ Jesus our Lord.

Faith is a powerful force in the Kingdom of God. We can easily say, *"We can't live without it!"* Without the power of faith in our lives, we would be spiritually useless. Why do I say this? The Bible declares, ***"Now without faith it is impossible to please God." (Hebrews 11:6, HCBS).*** Reading the eleventh chapter of ***Hebrews,*** we understand why faith is so important to the Old and New Testament believers. Each step we take in our Christian walk requires faith in God. Consider this question: Do you think you could live without having faith in God? I do not think we can, or rather, I know we cannot! The Lord has designed His children to believe in Him and trust His instructions that He has given us. Pleasing God is really believing, trusting, and obeying the Word that He has given us and the power to accomplish His desired will on earth.

When faith is securely established in our spirit, we begin to make inroads into the spiritual warfare that rises against the child of God. With any battle against our spiritual enemies, it is imperative that we have faith in the Word of the Lord and the God-given authority that we may fight and win. Will these fights be complicated? Some will be complicated and require lengthy battles. Some fights will take fasting and praying, while other fights may take a short period of time. It does not matter how short or long the battle is. If we trust the Lord and put the fight plan in the Hands of the Lord, we will win! I mention several spiritual battles that the children of God have to deal with. In each of these battles, it was imperative that the child of God depended on the Lord's power and strength to win the battles. It is imperative that these spiritual battle plans include the intel from Heaven. No prophet, priest, apostle, evangelist, teacher, or preacher can win the spiritual battles without the Lord. Likewise, as children of God during this twenty-first century, we need the Lord on our side and His directions to win every battle. Are you willing, determined, and able to win the ungodly battles that we will be faced with? Consider the voice of Caleb, one of the warriors of Israel during Moses' leadership. During Israel's conquest of the enemy, as they marched into battle, Caleb proclaimed, ***"Let us go up at once and take possession, for we are well able to overcome it."***

(Numbers 13:30, NKJV). Why did Caleb declare that they were well-able? He knew that the Lord was on their side and winning was not an option, it was a sure thing! When we fight with confidence, we will win. It does not matter what the negative circumstances may be. We will win because the Lord is on our side, and it is the Lord that's fighting for us!

The common question for most people, saved or unsaved, is this, *"At some point in my life, will I experience temptation, adversity, or trouble?"* If you have lived any time on this earth, you will say absolutely yes! A great percentage of humans on this earth will agree with the previous statement. Some will try to avoid the adversity that comes, but no matter how hard they try, adversity will come. It does not matter what age you are or your economic or spiritual status, you will experience temptation, adversity, and trouble. The answer I give you, which you may not want to hear, is simply this: you will experience these three things (temptation, adversity, and trouble). We have been told by teachers, preachers, lawyers, and doctors that all kinds of adversity will happen in our lifetime. We are faced with these three things from birth to death. We will have good and bad times, but in most cases, we face these times and continue to push on. However, if we are determined to see more good times than bad, we will make the wisest choice of all, and that is to invest our time in the Word of the Lord. It is the Word that helps us to overcome the multiplicity of problems that come our way. The Spirit of God that lives in us gives born-again believers the power and the ability to overcome any adversity that we may encounter. I am determined to receive the crown that the Lord has prepared for me! I talked extensively throughout this book about the people in the military and in athletics. In each case, if you abide and survive successfully, a crown awaits you.

Therefore, our fight to make it to the finish line must be *endurance*. Endurance must be the heartbeat of every Christian. We have been told a number of times that we will suffer percussion, hardships, and struggles in this life. Nevertheless, keep fighting and do not give up. The Lord will fight the battles for us as He gives us the correct instructions at the proper time. I am convinced He will! One of the enemy's greatest tactics is to cause doubt, worry, and unbelief in the child of God. Remember when Jesus was driven into the wilderness to be tempted by the devil? The devil's crafty and cunning words were, ***"If you are the Son of God?"*** How many times has someone tried to persuade you to do something wrong by using the *"If then"* quotation before you? It has happened to the best of us until we became wise to their tactics and moved on. Well, it is time for the child of God to move on and far away from all the ungodly tactics of the devil. The devil knows that he cannot win the war, so he tries to bring us down by causing us to be defeated in as many battles as he can to get us to turn back. But I have decided I'm not turning back. Something down in me keeps telling me to keep going. It does not matter the severity of the storm or the multitude of arrows that try to pierce our soul. I am a mighty **Christian** soldier and a professional **godly** athlete, and the words that will come out of my mouth are simply this, **"I have received power, and I will not be defeated!"** Take time to check out your spiritual manual and toolbox.

Everything you need is included in the box. Study what the Lord has placed in your spiritual manual and toolbox. You will find that you have everything you need to build up the Kingdom of God in you. Do not be afraid to fight, we will win and glorify the Name of the Lord Jesus. Keep fighting and working on the building that has been placed in our possession. A great reward awaits us! Therefore, work, fight, and live by faith, and we will see the glory of our Lord Jesus Christ in our lives!

I will speak of Your testimonies also before kings, And will not be ashamed. And I will delight myself in Your commandments, Which I love. My hands also I will lift up to Your commandments, Which I love, And I will meditate on Your statutes.

(Psalms 119:46-48, NKJV)

Testimonies

And this is the testimony: that God has given us eternal life, and this life is in His Son. He who has the Son has life; he who does not have the Son of God does not have life. These things I have written to you who believe in the name of the Son of God, that you may know that you have eternal life, and that you may continue to believe in the name of the Son of God.

(1 John 5:11-13, NKJV)

A Soldier's Story – Deacon Kenneth Willis

My experience in the military was challenging and rewarding. I spent 20 years in the United States Marine Corps as a Communicator. I was responsible for ensuring that Radio Communication between my Commander and his allies was always sufficient. As marines, we spend 13 weeks in boot camp at either Parris Island, South Carolina, or at Marine Corps Recruit Depot San Diego. I went to Parris Island at the age of eighteen after graduating high school. During my tenure at Parris Island, we were taught to change our mindset. For example, we learned to speak words like we, us, and our, and that would be the ideal terminology. We could no longer say I, me, or mine. It was all about becoming a team and not an individual. The training that I received in Boot Camp was the foundation of becoming a Marine (Basic Rifleman). After we completed boot camp, we were required to spend a month at Marine Combat Training (MCT). MCT is where a Marine learns how to operate various weapon systems, such as hand grenades, 50-caliber machine guns, etc. We also studied warfare and how to maneuver during conflict. We learned offensive and defensive tactics. These tactics consist of how to envelop the enemy, which meant to "surround." How to evade the enemy. After MCT, I was sent to a communication school in 29 Palms. I spent several months learning how to become a field radio operator. Using various radio emissions, such as Very High Frequency (VHF), High Frequency (HF), Ultra High Frequency (UHF), and satellite communications to name a few. If you ever watch a war movie and you see a young man with the radio attached to their back that was me. All this training was a precursor for what was to come in future conflicts. During my first few years of service, I was able to travel the world and see many places. We would deploy with Naval Ships, which is called the Marine Expeditionary Unit (MEU). We would take about 3000 Marines and board several naval vessels to cruise the ocean while doing various simulated operations. These operations prepared us for any conflict that would arise while we were out at sea. For instance, there was a scenario in Africa where we were supposed to evacuate the United States ambassador because of a civil war between two African nations. The rescue was ultimately called off; however, I remember the tension that we all felt. Although we had to stand down, we were well-trained, so we were confident in our ability to extract the diplomats, had the mission continued. Fast forward to the beginning of the Iraq war in 2003. All of my training in warfare and tactics came to a culmination. My unit was tasked to travel 400 miles into Iraq to establish a Forward Operating Base (FOB). The trek through Iraq took weeks. It was slow and methodical. I was the Battalion Executive officer's personal Radio Operator, which meant that I had to ensure that he received all updates from our allies and the Colonel from Kuwait. I never left the radio during that movement on the route to our FOB, where I ate and slept day and night near the radio. Our mission as a support squadron was to assist the Ground Combat Element (GCE) and Air Combat Element (ACE) with fuel and weaponry for aircraft and various ground vehicles and systems. I can vividly remember Cobra

Attack helicopters landing and receiving missiles and fuel from our location to engage the enemy. The enemy was so close that you could see the helicopters firing their missiles from the air. The mindset of the US military was to attack them from the air in order to achieve a strategic advantage. When one is attacked from the air, there is no covering from the opposing forces. This kind of pressure will eventually kill or force them to surrender, giving us the authority. There was another instance where we had to climb to the top of a ridge in preparation for an attack from the enemy. We took a defensive position with all our weapons in line, like you would see in an old war movie. We were in that position for hours expecting an attack; however, it never came from the direction that we were expecting it to. Just speculating, but I think they figured that we were able to defend that avenue of approach too easily. In 2005, I was tasked to escort the female Marines to all of the entry control points in the city of Fallujah, Iraq. The purpose was to allow the civilians back into the city after it was cleared out by US forces and their allies. The female Marines was used to search Iraqi woman as they reentered the city. This was done twice a day, which was a dangerous mission. We were constantly in a defensive posture because our orders were to "win the hearts and minds of the Iraqi people." This meant that we were to preserve life at all costs. The enemy did not have the same rules. We would report that the enemy was probing us. They would drive up to our convoys to see how we would react. My team was eventually taken from this assignment because we were too aggressive toward those probing us. We would shoot their vehicles with flares or threaten them with our machine guns. I did not want to lose anyone's life on my watch. A short while later, the same female Marines that I escorted were blown up by a Vehicle-Borne Improvised Explosive Device (VBIED). There were some survivors, but they were badly burnt and had to be discharged from the military. I have had to endure the stress and uncertainty of war on several occasions. The mind is the most vulnerable part of one's body in a war scenario. We are trained to stay in physical as well as mental shape, which allows us to continue to press beyond our normal limitations. As military men and women, we must learn we go through so many transitions in our careers. We go from one extreme to another. For example, we must focus on the mission at hand and still juggle the emotions of leaving our families for long periods of time, knowing that we may not return home. This takes a lot of mental endurance as well as physical. In order to keep my body going, I had to keep my mind clear of any negative thoughts. I had to believe that I was going to make it home.

Cancer Healing Testimony – Cheryl Cannida-Seaborn

In Oct 2013, I noticed a small pea-sized knot on my chest. I called my doctor, and he immediately had me get another mammogram. I did, and the radiologist told me right away I had a very rare triple-negative breast cancer and needed to see an oncologist right away! Then he asked me if he would call anyone for me. That's when Peace just fell over me as someone was already called! I was told I may not be able to get an appointment right away, but God! Not only did I get an appointment with one of the best oncologists in St Louis, but I got in the next day! I had the faith of a mustard seed. Every time they told me what they were going to do, I told them what God was going to do! The doctor told me this particular cancer usually only has a 5-year life expectancy! I told the Doctor, God's got me! As I went through chemo, they always checked my blood first. I told them there is power in the blood, and it's good! They always came back saying it was excellent. Next, they wanted a biopsy to see if the cancer was being spread so aggressively, which they said it did in all lymph nodes, but before surgery, they wanted another ultrasound. I told them what they saw was gone! I kept the faith. They did the ultrasound three times and found no cancer there! Then next was radiation. I was told the radiation would burn my skin. I said not this skin, as God's got me! Out of 32 radiations, not a single one burned! The doctor couldn't explain why...But God! And lastly, Physical therapy was advised because the side where the lumpectomy was performed would probably swell. I said it won't...God's got me! They did some measurements and said that the numbers were high and I would have to be wrapped the following week. I said no, as numbers will be down by then after prayers. By the next week, they followed up with additional measurements, and the technician came into the room saying she had never seen numbers drop down so fast and so low! God blocked it!!! Ten years later, I'm still here cancer-free! Faith is the substance of things hoped for and the evidence of things not seen! Faith is real, a word from God! I believe and accept the Word of the Lord. Thank you, Jesus!!

When God speaks – Harold Hunt

"It nurtured me, and it gave me the most satisfying years of my life."

I am writing this letter for several reasons, and I hope none of them are too trite. First, I would like to spare my friends, who may happen to read this, the usual clichés about being a good soldier. They were all kind enough to me, and I was not enough to them. Second, I would not want to be a party to the perpetuation of an image that is harmful and inaccurate: "Glory" is the most meaningless of concepts, and I feel that, in some cases, it is doubly damaging. And thirdly, I am quite simple and let God at all times have full authority over my life. In fact, HE has carried me so many times that I can't count because the numbers don't go that high. I loved the Army: it reared me, it nurtured me, and it gave me the most satisfying years of my life. Thanks to it, I have lived an entire lifetime in 30 years. It is only fitting that I should die and be buried in my uniform. We all have but one death to spend, and insofar as it can have any meaning, it finds it in the service of comrades-in-arms. I knew this when I first donned the uniform in 1961 and accepted it, but my love for the Army was great enough – and the promise that I would someday be able to serve all the ideals that meant anything to me through it was great enough – for me to accept this possibility as a part of a price which must be paid for all things of great value. If there is nothing worth dying for – in this sense – there is nothing worth living for. By the time I was 21 years old, I was sergeant sold on the doctrine of "Mission Accomplishment and Welfare of my Soldiers. On April 5, 1966, while on my second tour in Vietnam, something told me it was not going to be a good day, and sure enough, it wasn't. Death was all around us. However, on Easter Sunday, April 10, 1966, everything changed; I prayed for the fighting to stop; prayed for my newborn son, who was now one year old, and for my four younger brothers that they would never have to experience war. Also, I noted to my automatic rifleman, *"I don't ever remember meeting a true atheist. There were some that talked about not believing in the existence of God, but what I found in*

Vietnam when confronted with the possibility of death, there were no non-believers." April 21, 1966, God answered my prayer, for that was the day I was wounded. My exact words at that time were, "Oh my God, I've been hit, 'Medic'." Weak from my own blood loss, Medics rushed me to the aid station. Almost immediately, I was transferred to hospitals in Saigon, The Philippines, and Japan, where I recovered and returned to the States. Years later, I learned of the prayers of several of my family members at home for me at this time. They had gathered, not knowing of my situation, on the day of my injury to pray for my safety. "Without those prayers, I think I would have died on the battlefield. Up to this time, I didn't fear death because I felt God speaking to me: "I am sparing you and all of the soldiers you lead because your family members are praying for you at this time." During my third tour of duty in May 1969, I felt God spared me on other occasions during the war. In fact, during my third tour in 12 months, leading a platoon of 33 men, I didn't have even one soldier killed in action. Which goes to prove that I was not the only soldier who experienced God's protection. Of course, not every story has a happy ending. I'm convinced these soldiers entered the Vietnam War believing in God's providence. If they had lost their lives in service to their country, they would have accepted that as part of their duty and HIS plan for their lives. To those 945 soldiers of the 27th Infantry "Wolfhounds" Regiment, 25th Infantry Division, whose lives ended prematurely in the Vietnam War, you will always be remembered. The Army let me live in Germany with experiences in all of these places that others only dream about. I have toured the place where "Ann Franks" hid from the Nazis, gone to Paris, seen the nightlife in Frankfurt, danced to the Temptations, climbed the mountains where the Korean War was fought, and earned three academic degrees, including a master's degree in business management. I have known what it is like to be married to a fine and wonderful woman and to love her beyond bearing with the sure knowledge that she loves me; I have served at all echelons and have been a father, income-tax advisor, confessor, and leader for hundreds of men and women at one time. I have gone to Airborne and RECONDO School, and I have made parachute jumps with my younger brother and oldest son. I have written several books and articles that were published in Army Times, and I have studied mass rescue operations, compelled a book titled "The Challenges of Managing an Emergency at Sea," and sat on several international working groups with the International Maritime Federation and traveled to Europe, the Caribbean, and China. I have experienced all these things because I was in the Army or a government civilian servant. The Army is my life; it is such a part of what I was that what happened is the logical outcome of the life I lived. I never knew what it was to fail; I never knew what it was to be too old or too tired to do anything. I lived a full life in and out of the uniform, and it has exacted the price. It is only just.

Healing From Cancer – Sister Marian Sanders-Brookings

On April 10, 2010, I was diagnosed with stage-one breast cancer. I agreed to have a Lumpectomy and Chemotherapy. Before the Lumpectomy, God spoke to me and told me I needed prayer. I then contacted Bishop Jefferson, whom I have known for over 20 years. I asked him and his congregation to pray for me. When I went to my Chemotherapy appointment, the doctor informed me that Chemo could not begin because all the cancer was not removed. Therefore, I had a second Lumpectomy to remove the remaining cancer. Unfortunately, after the second surgery, cancer was still present. At this moment, I hit rock bottom. I could not pray for myself. I felt that I was not going to be healed because I was not worthy of being healed. This was the attack of the devil on my mind. During this time, I was mentally unstable, and my sister-in-law helped me find another doctor and had my left breast removed. I appreciate Bishop Jefferson because, through this uncertain time in my life, he was right beside me for each surgery, praying for my healing.

Now, looking back, I realize that the prayer God spoke of was not for healing but for the attack on my mind.

The Lord's Great Healing Power – Bishop John W. Frost

I send this testimony of the Lord's great healing power in my life to the praise of Almighty God. Several years ago, while I was still in the Washington DC area and pastoring, I started having great pain in my lower back. I learned from the doctors that I had two discs fused in the lower part of my spine. When the pain hit me in the area, I would scream at the top of my voice. I cried out to the Lord to heal me and take away the pain. After doing this several times, I heard the voice of the Lord telling me that His grace was sufficient for me. Healing never came, but when I cried out "Your Grace Lord," the pain would ease. After a while, the pain was there, but I did not pay attention to it. Coming forward in time, on Sunday, September 3, 2023, I got up and dressed for church. I heard the voice of the Lord telling me that He would heal my lower back pain. When I got to church, there came a prophet who had come to worship with us. I had not met or talked to him in the past. After the service, he asked me to come up into the pulpit area. He spoke to me about support for the pastor and that I was to stand behind him and beside him to lift the pastor's arm. Then he took me aside from standing behind the pastor and said these words to me. He said that the Lord said that the pain I had from the fused disc was healed. At that moment, the pain was gone. Great is the faithfulness of our God, and His grace is sufficient for us and will sustain us until His healing takes place. He healed me, and I have not felt that pain since. This is one of the many areas of healing that I have received from the Lord. Thank you for giving me the opportunity to share this with you and whomever you wish to share with.

Bishop John W. Frost

The Day I Died – A Testimony By James Williams

I am a retired U.S. Army Intelligence Officer with 20 years of active-duty service to the greatest nation on earth. I provided sensitive support to elite Department of Defense (DoD) units, operations, programs, and activities. I have conducted high-threat operations and low-vis activities in various areas of responsibility (AORs) from the Cold War Germany, Southwest Asia, Iraq, Bosnia, Kosovo, Afghanistan, and others. Risking one's life and experiencing the horrors of war is simply a condition of the job.

Unfortunately, I have suffered the loss of more comrades to suicide than combat. According to the American Psychological Association, approximately 17 Veterans commit suicide each day, and the current suicide rate for Veterans is 1.5 times higher than for non-veterans. As a God-fearing soldier who had achieved things that only a chosen few can ever know, I, too, almost fell victim to suicide. It wasn't the trauma of battle or conflicts with the morality war that led me to that decision. It was the deceptive notion of hopelessness after returning home to a mundane life, an ungrateful nation, and a looming divorce. I experienced depression from an overwhelming, flawed belief that my life lacked purpose and routinely asked myself, "What was it all for?" This is the age-old question of purpose and impact, which God alone can provide that I was searching for. At that critical moment, I failed to realize that my service, albeit noteworthy, was anchored merely on the doctrine of works rather than faith in Christ Jesus. God tested my work that evening while lying on the floor, hidden from a worldview in shame. I'm now reminded of what the scripture says, ***"For God will bring every work into judgment, including every secret thing, whether good or evil (Ecclesiastes 12:14)."*** I was blinded by the bias of my own carnal perspective, which birthed faulty conclusions about my life. I was wrong, and God had something to say about it.

The day I died, I realized the Power of the Lord. The day I died, there was silence, no sounds of war, no screaming, no explosions, nor the clap of small arms fire. I died at home on the bathroom floor—alone. As I laid on the bathroom floor wallowing in self-pity and perceived despair, I planned to end my life. Before the final act, the voice of the Lord boomed in my ears. The Lord said, "You better think **long** and **hard** about what you're planning to do!" Now, wait a minute, this was not the sweet and comforting voice that one would expect to receive at the lowest point of their life. Instead, the imposing tone sent shivers to the depths of my soul, and an overwhelming sense of paralyzing fear seized the very breath from my quivering lips. I was then engulfed in a keen spiritual understanding of the eternal implications of my decision. The Shock and Awe of the Persian Gulf War was nothing compared to the Anger of the Lord. I was reminded that, ***"It is a fearful thing to fall into the hands of the living God (Hebrews 10:31)."***

My response was as pitiful as my self-imposed position. However, I did what all trained intelligence professionals do in the face of intense scrutiny or interrogation—shift the blame, make counter-accusations, or manipulate the situation to attribute it to some sort of misunderstanding. Okay, here it goes—I proceeded with a well-articulated and ardent plea to illustrate and quantify my perceived misfortune coupled with examples and the resultant impacts, all designed to move even the most obstinate judge or jury. Well, that didn't work. In response, the Lord cross-examined me sternly, saying, "Are you qualified to sit on My throne?" I liken this encounter to when God asked Job, ***"Where were you when I laid the foundations of the earth? Tell Me if you have understanding (Job 38:8)."***

In short, God wasn't having it. Without debate, I am not God and definitively unqualified to be in His presence, let alone His throne. Notwithstanding the sheer audacity and futility of my actions, God cannot be manipulated. There are no excuses, no personal justifications, nor any other response, no matter how seemingly eloquent or profound, to explain away our sin. The Bible says for it is written: ***"As I live, says the Lord, Every knee shall bow to Me, And every tongue shall confess to God. So then each of us shall***

give account of himself to God." (Romans 14:11-12)."

I knew that I clearly lost that round and that it was now time to man up and own it. I knew that I should take responsibility for my actions, but instead, I chose the cowardly approach—I threw a temper tantrum (aka the "Why me?" defense). My rebuttal started with, "Lord, I tried everything and failed anyway. What's the point, I have nothing left. I'm tired of fighting, and I'm empty." Nope, that didn't work either, as evidenced by God's sudden silence. I don't know if Jesus was listening patiently to my rant or just letting me cry it out like a misbehaving two-year child. In the end, I simply conceded defeat to His wisdom, authority, and power. Since my loss was evident, I switched to a closing plea for utter mercy. In frustration, I said, "Lord, can't you see that I just died here!" The Divine response thundered once again, and the frank tone shook me to my core. Jesus simply responded, "Now rise and live!" This was a command, not an option. Jesus then explained that death to this world and to my carnal ways was the beginning of a fulfilling life in

Him. I had to first die in order to live. Once you die to yourself, Christ becomes your purpose, and as you walk with Him, your impact becomes evident. At that moment, I surrendered and died, and then God chose to make me Rise and Live.

Yes, I have experienced and overcome many more tragedies over the years and will undoubtedly experience more. It's life and struggles are just an environmental condition. The difference is that I know ***that "Many are the afflictions of the righteous, But the Lord delivers him out of them all." (Psalm 34:19).*** It's been more than 20 years since my death, and my new life has exceeded all expectations and dreams. Truly the Lord can do,***…exceedingly abundantly above all that we ask or think, according to the power that works in us…" (Ephesians 3:20)***. I can firmly attest to the implicit power of God and the abundance of His Grace and Mercy to deliver you from all harm, including self-harm. Now, as a minister of this great gospel of Christ Jesus, I know that God uses shadow warriors to usher His light into the darkest situations.

The Lord Saved My Life – Brother Joe Jones

After being diagnosed with Prostate cancer about twelve years ago and going through surgery, I remember the doctors and nurses asking, how was I so calm? I could not explain it at that time, but after the surgery was over, I had time to reflect. All I could think of, it had to be GOD. It was GOD comforting me and letting me know He was with me as He had promised. I didn't have to worry about what was going to happen. Because of His Love for me, He touched and healed me.

When GOD heals you, and you receive and are able to testify, I had to give the Lord all the glory for what he did for me. Because of his Love for me, he touched and healed me. I discovered some tremendous things about the power of God's healing. I learned that attitude and prayer affect healing.

I have been cancer-free since 2011, and I have to give God all the praise and glory. I have to give my entire family, especially my wife and friends, thanks for their love and support. The way they stuck by me, showing so much love, is priceless. Having cancer has taught me to love each and every day like it is my last day. No matter what life throws at me, God will be in it with me. I learned going through this journey that doctors don't heal us, only God can do that!

God is the head of my life..........

The MagNiFiGreat Miracle Experience Prophetess Gloria Wynn Lilly

"Moreover, King Hezekiah officials ordered the Levites to sing praises to the Lord with the words of David and Asaph the seer, so they sang praises with joy and worshipped!" 2 Chronicles 29:30

In 2009, I had a massive headache, and God said to go to the ER. I had a large brain tumor! I was referred to a Neurologist. He ordered an MRI and found that I also had a smaller tumor between my eyes. Since I wasn't in any immediate danger, we would watch the tumor and, if needed, call the office or go to the ER. In 2010, 2011, and 2012, I had a massive headache, went to ER, got more pain meds, went home, pain subsided. Then, on April 28, 2013, while introducing the speaker for Women's Day, I began stuttering, and it got worse and worse. I turned to the Guest Speaker and said, "Something's wrong, I can't say the words." She got up, prayed for me, and I went back to the organ to play the closing song. When I arrived at the ER, my neurosurgeon declared, "It looks like this large tumor has gotten out of hand!" Had more tests done, and they called in a top neurosurgeon to assess the situation. He said I had a Cancerous Hemangiopericytoma, a tumor like a parasite that likes to grow in any organ in your body. Neurosurgeons said it would probably take 3 hours to operate. The next day, they said the operation would take 8.5 hours, four plus hours for two surgeons. They didn't know when or if I would wake up. I could have complications during or after surgery, or I could die. There were NO guarantees. I did not panic. I went home, prayed, and prepared for surgery.

On May 17, 2013, at 6 a.m., I started chanting, *"There is Power in the Blood of Jesus,"* until 1 got to the hospital. There were 15 family members and friends praying in the Healing Garden at the University of MD, Baltimore, MD. "God told me He was going to give me a "MagNiFiGreat Miracle." *("MagNiFi"—something beyond the norm, man couldn't do, never been done before: "supernatural, amazing, and extraordinary, "Great," gigantic, enormous, and profound)*. I believed Him. My surgery was an absolute success! "I woke up in my right mind praising God, and I knew I was Alive!" Still a little groggy, surgeons said I had to have an MRI to see what the surgery had done. While in the room, "l saw a vision of a light beaming directly upon me appearing to be an image of Christ in a *White robe.* I was lying down, looking up to Jesus. He spoke to me, saying, "I am the Lord thy God that is healing you." At that moment, I felt his glorious presence and bright light surrounding me like a "halo", illuminating me! Oh my God! "All I know is I was in the presence of a Heavenly Host!"

I was so grateful that I wrote a new song on the 3rd day in the hospital entitled, "Expect the MagNiFiGreat Miracle!!" HALLELUJAH! God still wasn't done. I needed 2-3 months of radiation therapy to kill the cancer. Every day in therapy, the team played worship songs, and I basked in his miraculous healing power. My miraculous recovery occurred from 2013 to 2023. The large humorous tumor did not return. The smaller tumor between the eyes stabilized. Major stroke stabilized in 2018. Mini strokes stabilized in 2019. Small pituitary tumors stabilized. Praise God for my spectacular medical staff of Neurosurgeons, Oncologists, Specialists, Nurses, Housekeepers, and Security teams at the University of Maryland, who performed an outstanding job. "God Chose Me to Share my Story For His Glory, and He wants me to tell the World that He Still Performs Miracles every day, and He wants to Perform a Miracle in your Life Today. "If you need a Miracle, all you have to do is ask God!" Over the last 30 years, I was able to pray and witness God's saving power in the lives of my Mother, Brothers, adopted Mother, and saints; ultimately, God saved my Life!

References

Amplified Bible (AMP), (1987), Publisher: Zondervan Publishing House; Amazon.com: Books

Bacon, Frances Sir, Meditations Sacrae (1597), the saying: "knowledge itself is power", Having and sharing knowledge is the cornerstone of reputation and influence, and therefore power; all achievements e, Cyanate from this.

Bible by Logos, Logos Bible Software by Faith Life Corporation, Wikipedia, 1992, C^{++}, C#, Logos Research Systems Inc.,

Eastern Standard Version (ESV), (2001), Amazom.com

Gaither, Bill; Gaither Vocal Band; ***"Get All Excited"*** Christian, Gospel, Country, Pop; Lyric Band, November 1, 2004, March 5, 2022

Gilcrest, Earl; 1949, Retired Engineering Supervisor with General Motors; Flint, MI

Google; 1998, Python C, C^{++} Search engine

Harmon, Matthew S., Professor of New Testament studies Biblical Theology, Wheaton College, Wheaton, Illinois, M.Div. 2003 Trinity Evangelical Divinity School, Deerfield, Illinois B.S. 1995 Communication, Ohio University, Athens, Ohio.

Hawkins, Bishop George W., Sermon: (2023) *"The Blessing from Bitten Hands."*

Holman Christian Standard Bible (HCSB); (1999, 2017); WORDsearch and Bible Explorer software.

Jefferson, E.H., (1949) Emmanuel Temple Church, Columbia, MD; Loyola University Maryland M.A. 2009; University of Missouri St. Louis, B.S., 1989

Kenyon, E. W., Two Kinds of Faith, 9781577700081 - Amazon.com"

King James Version (KJV); (1611, 1769)

New International Version (NIV) (1995); 1973, 1978, and 1983

New King James Version (NKJV) Copyright 1979, 1980, 1982, Thomas Nelson, Inc.

New Living Translation (NLT); (2008); Bible Gateway, Zondervan; HarperCollins Christian Publishing

Made in the USA
Monee, IL
04 November 2026

67383411R00105